154

Advances in Polymer Science

Springer-Verlag Berlin Heidelberg GmbH

Polymer Physics and Engineering

With contributions by
M.D. Barnes, K. Fukui, K. Kaji, T. Kanaya, D.W. Noid,
J.U. Otaigbe, V.N. Pokrovskii, B.G. Sumpter

 Springer

This series presents critical reviews of the present and future trends in polymer and biopolymer science including chemistry, physical chemistry, physics and materials science. It is addressed to all scientists at universities and in industry who wish to keep abreast of advances in the topics covered.

As a rule, contributions are specially commissioned. The editors and publishers will, however, always be pleased to receive suggestions and supplementary information. Papers are accepted for „Advances in Polymer Science" in English.

In references Advances in Polymer Science is abbreviated Adv. Polym. Sci. and is cited as a journal.

Springer APS home page: http://link.springer.de/series/aps/ or
http://link.springer-ny.com/series/aps
Springer-Verlag home page: http://www.springer.de

ISSN 0065-3195

ISBN 978-3-662-14674-3   ISBN 978-3-540-44484-8 (eBook)
DOI 10.1007/978-3-540-44484-8

Library of Congress Catalog Card Number 61642

© Springer-Verlag Berlin Heidelberg 2001
Originally published by Springer-Verlag Berlin Heidelberg New York in 2001
Softcover reprint of the hardcover 1st edition 2001

The use of registered names, trademarks, etc. in this publication does not imply, even in the absence of a specific statement, that such names are exempt from the relevant protective laws and regulations and therefore free for general use.

Typesetting: Data conversion by MEDIO, Berlin
Cover: MEDIO, Berlin
Printed on acid-free paper   SPIN: 10706234   02/3020hu - 5 4 3 2 1 0

Editorial Board

Advances in Polymer Science
Now Also Available Electronically

For all customers with a standing order for Advances in Polymer Science we offer the electronic form via LINK free of charge. Please contact your librarian who can receive a password for free access to the full articles. By registration at:

http://link.springer.de/series/aps/reg_form.htm

If you do not have a standing order you can nevertheless browse through the table of contents of the volumes and the abstracts of each article at:

http://link.springer.de/series/aps/

There you will find also information about the

– Editorial Bord
– Aims and Scope
– Instructions for Authors

Contents

Generation, Characterization, and Modeling of Polymer Micro- and Nano-Particles

Joshua U. Otaigbe[1], Michael D. Barnes[2], Kazuhiko Fukui[2], Bobby G. Sumpter[2], Donald W. Noid[2]

[1] Department of Materials Science and Engineering, and of Chemical Engineering, Iowa State University of Science and Technology, Ames IA 50011, USA
e-mail: otaigbe@iastate.edu
[2] Chemical and Analytical Sciences Division, Mail Stop 6142, Oak Ridge National Laboratory, Oak Ridge, Tennessee 37831, USA

Polymer micro- and nano-particles are important in many technological applications, including polymer blends or alloys, biomaterials for drug delivery systems, electro-optic and luminescent devices, and polymer powder impregnation of inorganic fibers in composites. They are also critical in polymer-supported heterogeneous catalysis. This article reviews recent progress in experimental and simulation methods for generating, characterizing, and modeling polymer micro- and nano-particles in a number of polymer and polymer blend systems. A description of the use of gas atomization (of melts) and microdroplet (solution) approaches to generation and characterization of spherical polymer powders and microparticles represents their unique applications, giving the non-specialist reader a comprehensive overview. Using novel instrumentation developed for probing single fluorescent molecules in submicrometer droplets, it is demonstrated that polymer particles of nearly arbitrary size and composition can be made with uniform size dispersion. This interesting finding is ascribed to new dynamic behavior, which emerges when polymers are confined in a small droplet of solution the size of a molecule or molecular aggregates. Solvent evaporation takes place on a time scale short enough to frustrate phase separation, producing dry pure polymer or polymer blend microparticles that have tunable properties and that are homogeneous within molecular dimensions. In addition, it shows how a number of optical methodologies such as Fraunhofer diffraction can be used to probe polymer particles immobilized on two-dimensional substrates or levitated in space using a three-dimensional quadrupole (Paul) trap. Unlike conventional methods such as electron-beam microscopy, the optical diffraction methods provide a unique look inside a polymer particle in a measurement time scale of a few milliseconds, making it attractive to in-line production applications. In particular, it shows that it is possible to use computational neural networks, extensive classical trajectory calculations (i.e., classical molecular dynamics methods) in conjunction with experiments to gain deeper insights into the structure and properties of the polymer microparticles. Overall, it is possible to use the new understanding of phase separation to produce a number of useful, scientifically interesting homogeneous polymer blends from bulk-immiscible components in solution. Additionally, this new knowledge provides useful guidelines for future experimental studies and theory development of polymer and polymer blend micro- and nano-particles, which are not widely studied.

Keywords. Polymers, Micro- and nano-particles, Particle characterization, Microdroplet, Gas atomization, Molecular dynamics and neural network modeling, Optical diffraction-based probes

1
Introduction

Polymer particles of different shapes and sizes are critical in numerous applications, including polymer blends or alloys [1], polymer powder spray coating [2], and polymer powder impregnation of inorganic fibers in composites [2] and in polymer-supported heterogeneous catalysis [3–6]. Recently, significant commercial and scientific attention has been focused on multi-component polymer systems as a means for producing new materials on the micrometer and nanometer scale. Composite polymer particles or polymer alloys with specifically tailored properties could find many novel uses in such fields as electro-optic and luminescent devices [7–8], thermoplastics and conducting materials [9], hybrid inorganic-organic polymer alloys, and polymer-supported heterogeneous catalysis.

The feasibility of using gas atomization processing to prepare micrometer-sized spherical particles directly from molten polyolefins was recently demonstrated [10–12]. Gas atomization processing of polyolefins shows promise for efficiently producing large amounts of polymer powders with particle sizes ranging from 25–200 micrometer. However, it is felt that generating monodisperse polymer particles and polymer blend microparticles from bulk-immiscible components may lead to new polymer alloys on the nanometer and micrometer scale with specifically tailored material, electrical, and optical properties. The microparticle generator will offer a new tool for studying multi-component polymer blend systems in femtoliter and attoliter volumes, where high surface area-to-volume ratios play a significant role in phase separation dynamics.

Recently, a new experimental method for producing polymer particles with arbitrary size and composition was reported [13]. The method is based on using droplet-on-demand generation to create a small drop consisting of a polymer mixture in some solvent. As the solvent evaporates, a very precise polymer particle is produced. These particles in the nano- and micrometer size range will provide many unique properties due to size reduction to the point where critical length scales of physical phenomena become comparable to or larger than the size of the structure. Applications of such particles take advantage of high surface area and confinement effects, leading to interesting nano-structures with different properties that cannot be produced using conventional methods. Clearly, such changes offer an extraordinary potential for developing new materials in the form of bulk, composites, and blends that can be used for coatings, opto-electronic components, magnetic media, ceramics and special metals, micro-or nano-manufacturing, and bioengineering. The key to beneficially exploiting these interesting materials and technology is a detailed understanding of the connection of microparticle technology to atomic and molecular origins of the process. It is felt that microparticle technology may lead to rational approaches and complementary theories of the unresolved problem of phase-separation in polymer blends, as discussed in Sect. 4.

While great success has been achieved in understanding the kinetics of phase-separation in polymeric systems [14–21], finding the conditions that frustrate phase separation in polymer blends remains an elusive goal. To develop materials for new applications, progress in approaches that exploit phase-separation mechanisms as well as approaches that frustrate it are needed. Bates and coworkers reported an approach for generating such unusual structures. Their approach relies on exploiting microphase separation driven by chemical incompatibilities between different blocks that make up block copolymer molecules [22]. Similar approaches on the role of metastable states in polymer phase transitions can be found in a book by Hamley [23] and in the classic, elegant review by Cheng and Keller [21]. Comparatively, however, the microdroplet and gas atomization approaches to generating and characterizing polymers, polymer composites, and alloys on the micron and submicron scale have not been investigated to the same extent despite their theoretical and practical importance.

The present article is intended to discuss state-of-the art experimental approaches to generation, characterization, and theoretical analysis of polymer microparticles. Chemical synthetic routes (such as dispersion and macromonomer techniques) to generating polymer microparticles have been recently reviewed [24–25]. Therefore, we intend here to mention a few examples of chemical synthetic routes to polymer microparticle generation and describe only the very recent important developments in polymer microparticle science and technology. Generation, characterization of polymer powders and microparticles by use of gas atomization and microdroplet approaches will be described in some detail to give the non-specialist reader a comprehensive overview and to represent their unique applications. Using instrumentation developed for probing single fluorescent molecules in submicrometer droplets, it will be demonstrated that polymer particles of nearly arbitrary size and composition can be made with a size dispersion of typically approximately 1%. Extensive classical trajectory calculations (using classical molecular dynamics methods) will be used to gain deeper insights into the structure and properties of the polymer microparticles.

Some comprehensive textbooks covering earlier engineering aspects of polymer powder technology are available [1–2].

2
Survey of Polymer Powder and Microparticle Generation Techniques

Fukui et al. [26–27] and Kung et al. [28] reported molecular dynamics simulation of nanometer scale polyethylene (PE) particles generated with≤12,000 atoms to gain insight into some thermodynamic properties of ultra-fine polymer powders. They found that simulation of the molar volume and total energy as a function of temperature resulted in values for melting point, glass transition, and heat capacity that were dependent on the size of the polymer particles. The simulation predicted a reduction of the melting point on comparison to the bulk system. Further, they found that the ratio of surface atoms for the polymer particles was extremely large and surface-free energy was dramatically dependent on the size of the particles. Using a piezoelectric driven droplet generator, Kung et al. [28] produced ultra-fine polymer particles for both water-soluble and water-insoluble polymers. Spherical particles of PEG, PVA, and PVC ranging in size from 300 nm to 10 μm with approximately 1% monodispersity were generated by varying the solution concentration and orifice size [28–29].

Polymer-assisted synthesis of ultra-fine metal or semiconductor particles in solution, to arrange the particles in a matrix in well-controlled fashion, for possible electro-optical applications was recently reported by Scholz et al. [30]. This method allows synthesis of gold nanoparticles by reducing gold chloride. Gold was found to be present in the form of complexes, to the polyelectrolytes, or in form of nanoparticles or larger aggregates, depending on the relative concentrations of the reagents and polymer molecular weight. Using a relaxative auto-dispersion (RAD) process, consisting of preparing an amorphous polymer, vapor depositing the metal onto the polymer, and heat treating Deki et al. [31] obtained ultra-fine metal particles (5–10 nm) embedded in a nylon 11 matrix. An amorphous nylon 11 was prepared by quenching rapidly from the melt by a splat method and high-speed twin roller method. The resulting composites showed colors characteristic of the constituent colloidal metals. The RAD process was used by Masui et al. [32] to prepare ultra-fine particles of cerium oxide dispersed in polymer thin films. Selected-area electron-diffraction patterns of the particles were completely indexed as those of cerium (IV) oxide with cubic fluorite structure, and the lattice constant calculated from the radii of the Debye-Sherrer rings was 5.41 Å. Noguchi et al. [33] developed a similar RAD process for the uniform dispersion of ultra-fine metal particles in a uniform, isolated condition in nylon 11 films. The composite films could be used as electrical, magnetic, optical, and chemical materials.

Hifumi et al. [34] applied ultra-fine platinum particles protected by poly(Me acrylate-co-N-vinyl-2-pyrrolidone) to immunological detection of methamphetamine (MA). The polymer-protected ultra-fine particles chemically bound anti-methamphetamine monoclonal antibody to their surfaces. The antibody-fixed particles behaved like an antibody in the immunoreaction, making it possible to detect the MA to a concentration of ca. 10 ng/mL. Uda et al described a similar application of polymer-protected ultra-fine platinum particles to the immunological detection of human serum albumin [35]. Tamai et al. [36] showed that ultra-fine metal particles could be immobilized on fine copolymer particles that were produced by reducing copolymer particles-metal ion complexes. Transmission electron microscopy and X-ray diffraction were used to confirm that ultra-fine noble metal particles with a diameter below 10 nm were formed and uniformly immobilized on the surface of copolymer particles.

Chen et al. [37] studied a series of complexes of styrene-4-vinylpyridine co-polymers or poly(4-vinylpyridine) and transition metal chlorides. The transition metal-polymer complexes were used to prepare ultra-fine metallic particles dispersed in a polymer matrix by chemical reduction. Upon reduction, the metal ions were transformed into the corresponding nanometer scale metal particles with the protective polymers preventing the metal particles from oxidation and excessive aggregation. Ohtaki et al. reported the effects of polymer support on the substrate selectivity of covalently immobilized ultra-fine rhodium particles as a catalyst for olefin hydrogenation [38].

Synthesis of cobalt nanoparticles in a polystyrene/triphenylphosphine polymer matrix was reported by Leslie-Pelecky et al. [39] who showed that magnetic properties of the as-synthesized nanocomposites ranged from super-paramagnetic to ferromagnetic with coercivities on the order of 130 Oe. They found that annealing in vacuum produced coercivities of up to 600 Oe and remanence ratios of up to 0.4.

A method for cationic polymerization of bulk liquid monomers that incorporates ultra-fine metal particles into polymer matrices has been reported by El-Shall and Slack [40]. In this technique, pulsed-laser vaporization was used to initiate polymerization in the gas phase and cationic oligomers were injected into the monomer liquid phase, forming metal clusters that rapidly condensed to form ultra-fine metal particles which provided catalytic centers for the propagation of the polymer chains. Using two kinds of ultra-fine CdS particle-dispersed acrylonitrile-styrene polymer films, Hayashi et al. [41] showed that narrowing particle size distribution enhanced the values of third-order nonlinear optical susceptibility due to the decreasing overlap between the higher energy levels and the lowest energy level in the UV-visible adsorption spectra.

Spherical ultra-fine polymer particles (10–35 nm) were prepared by chemical polymerization of aniline in an inverse water-in-oil microemulsion [42]. This microemulsion method provided a denser, more uniform and compact film of higher condensation than that produced in an aqueous medium [16].

Titania and silica glass thin films containing Au and Pt ultra-fine particles can be prepared under ambient temperature and pressure using a sol-gel process described elsewhere [43]. The sol-gel processing conditions can be adjusted to give Au particles with the smallest mean diameter of 2.8 nm, leading to films showing plasma absorption at 530–620 nm and absorption coefficients up to 1.0×10^3 cm^{-1}. Tsubokawa and Kogure [44] reported on surface grafting of polymers onto inorganic ultra-fine particles. They prepared polymer-grafted ultra-fine titania, silica, and ferrite particles by reacting functional polymers containing terminal hydroxyl or amino groups, such as poly(propylene glycol) and diamine-terminated di-Me siloxane with acid anhydride groups on these ultra-fine inorganic particles [44]. The percentage of grafting increased with the increasing acid anhydride group content of the surface. The polymer-grafted ultra-fine particles gave a stable colloidal dispersion in organic solvents [44].

Neinhaus et al. [45] used simple model systems such as ultra-fine Fe(OH)$_3$ particles (approximately 30 Å in diameter) with pronounced dynamical features to get deeper insight into the complicated dynamics of large molecular networks such as the cation exchanger Dowex 50 W solvated with 60 wt. % aqueous sucrose. For such systems, they observed broad diffusional lines of varying widths in the Mossbauer spectra recorded at 80–350 K, proving the bounded nature of the diffusion [45].

The preceding section shows that relatively little information on polymer microparticles is available in the scientific journal literature, which is surprising considering their potential technological importance. Comparatively, much work on novel production methods and technological applications of this special class of materials has been disclosed in the patent literature by our gifted industrial colleagues. To appreciate the useful application in the design of polymer microparticles, we intend in the next section to outline briefly and describe only the recent important developments in the field.

2.1
Survey of Patent Literature

Table 1 shows a listing of patents on production methods and applications of polymer micro and nanoparticles. Recently, Otaigbe et al [46] described a method for making polymer microparticles, such as spherical powder and whiskers (a whisker is defined here as a polymer microfiber with <100 µm in length and a diameter of <10 µm). The method involves melting a polymer under conditions that avoid thermal degradation of the polymer, atomizing the melt in a special gas atomization nozzle assembly in a manner to form atomized droplets, and cooling the droplets to form polymer microparticles. The gas atomization parameters can be controlled to produce polymer microparticles with desired particle shape, size and distribution. The dynamics of the gas atomization processing method and the properties of the product polymer microparticles are described in Sect. 3. Handyside and Morgan [47] used rotary, two-fluid, or ultrasonic wave melt atomization processes to prepare thermosetting polymer powder compositions suitable for powder coating processes. The thermosetting resin may consist of polyester or epoxy polymer containing a curing agent and one or more coloring agents. The melt-atomized powder is characterized by improved particle size distribution and by a generally rounded particle shape [47].

Noid et al. [48] used a new device called a microdroplets-on-demand generator (MODG) to produce polymer micro- and nano-particles from solution. The proof of concept was demonstrated using poly(ethylene glycol) microparticles generated with the MODG and captured in a microparticle levitation device. The potential application of the MODG in materials science and technology is described later.

Aoki et al. [49] developed a method for making aqueous dispersions of ultrafine cross-linked diallyl phthalate polymer particles with average diameter 10–300 nm by polymerizing aqueous solutions containing up to 15% diallyl phthalates in the presence of 7–30% (on diallyl phthalate) water-soluble polymerization initiators without the presence of surfactants. The dispersions are useful as

Table 1. Examples of patents on production methods and applications of polymer micro and nanoparticles

Method	Microparticle	Application	Ref.
Gas atomization	Polyolefins and blends, and Epoxies and PET	Polymer microparticles	46, 47
Microdroplet	PEG, PS, EVA, blends	Polymer micro and nanoparticles	48
Dispersion polymerization	Diallyl phthalate polymer	Rubber and plastic modifiers	49
Free radical polymerization	PMMA	Good purity powders for medical uses	50
Melt blending	Polyolefins	Adsorbents and antiblocking agents	51
Mixing and ionization	Metal salt and PET	Electrically conducting polymers	52
Ultrasonic grinding	PTFE	Molding compounds	53
Macromonomer	Poly (N-vinylaceta-mide-g-styrene)	Antistatic and abrasion-resistant thin films	54
Dispersion	Fe and poly acrylates	Printing ink jet inks and products	55
	PMMA		56
	Resins and pigments		57, 58
Microphase separation	Poly(2-vinyl pyridine)	Heterogeneous catalysts	59
Grinding	Ca(OH)$_2$/Polyolefins	Molding compounds for batteries	60
Dispersion polymerization	Fe and organic polymer	Magnetic permeable composites	61
Dispersion	SiO$_2$ and PVC	Far-IR radiation emitting films	62
Collision crushing	PE-PP blend	Thin film formers	63
Solution spreading	PS coated on chrome	Composite thin films	64
Dispersion	Au, Pt/organic polymer	Colorant for transparent films	65
Evaporatiion and precipitation	Pigments and polyolefins	electrostatic photog. image developing agents	66
Dispersion polymerization	Fe and organic polymer	Superparamagnetic composites	67
Graft copoly-merization	Vinyl polymers	Acid-rain and soil resistant topcoats	68
Spreading and cur-ing	Polyisocyanates/acrylates	Water-resistant coating	70

Table 1. continuing

Dispersion polymerization	Polystyrene	Coatings, adhesives, and medical uses	71
Emulsion polymerization	Vinyl polymers	Alkali and freeze-thaw cycle resistant films	69
Emulsion polymerization	Polyacrylates	Functional coatings and water-resistant adhesives	72–74
Emulsion polymerization	Polyacrylates, etc.	Polymer latexes	75–78
Dispersion	SiO$_2$ and polyacrylics	Artificial stones	79
Chalcogenation reaction	Polyvinylpyrrolidone and metal compounds	Optical filters and nonlinear optical materials	80
Dispersion	Cuprous halide-dispersed PMMA	Optical filters and nonlinear optical materials	81

modifiers for rubbers and plastics. Organic monomers such as MMA or oligomers, optionally containing polymerization initiators, can be sublimed into reactors in vacuo in inert atmospheres and irradiated with UV light to give ultra-fine PMMA particles (3000 to 5000 Å particle sizes) with good purity [50]. A method for manufacturing spherical and uniform-size polyolefin ultra-fine particles is reported by Yamazaki and Takebe [51]. In this method, the ultra-fine particles (approximately 1.5 μm particle size), useful for supports of absorbents, anti-blocking agents, etc., are prepared by blending polyolefins in liquid organic compounds, melting the blends, cooling to form spherical polyolefin particles, and removing the organic compounds by extraction. In another method [52], electrically conductive polymer ultra-fine particles (0.5 to 10 μm particle size) are prepared by mixing an organic solvent solution of a metal salt with another organic solvent solution of a thermoplastic polymer, cooling or pouring this mixed solution into water or a poor solvent of the thermoplastic polymer to separate the metal salt-containing thermoplastic polymer particles, and conducting the precipitation of metal from the metal salt by the difference of ionization or by the addition of a reducing agent. Powdered poly(tetrafluoroethylene) (PTFE) with a specific surface area of 2 to 4 m^2/g and a low pressure molding coefficient of 20–150 is ultrasonically ground to give powdered PTFE that has a specific surface area of 4 to 9 m^2/g and a low pressure molding coefficient of <20 [53]. The PTFE powders are useful for moldings having high density (e.g., 2.1872) and good surface smoothness.

Polymeric ultra-fine particle-adsorbed structures with antistatic, low-friction, and abrasion-resistant properties were prepared by Akaishi, et al. [54]. The structures comprise various substrates laminated with charged polymeric thin films on which charged polymer ultra-fine particles, prepared by a macromonomer method are adsorbed. The structures are manufactured by immersing charged polymeric thin film-laminated substrates into a solution containing dispersed charged polymeric ultra-fine particles prepared by the macromonomer method to adsorb the particles on the thin films. As an example of this invention by Akaishi et al. [54], a quartz oscillator microbalance as a substrate was alternately immersed 10 times into solutions of polyallylamine hydrochloride and Na styrenesulfonate homopolymer to form multi-layer films having the homopolymer layer as the outermost layer. This was immersed in a solution containing dispersed N-vinylacetamide-grafted styrene polymer ultra-fine particles in the presence of NaCl to give a polymeric particle-adsorbed structure in which the adsorption of particles depended on the concentration of NaCl. For a review of the macromonomer method for preparing polymer particles, the reader is referred to the classic, elegant review by Ito and co-workers [24–25].

Printing inks and products made from them employ polymer and inorganic ultra-fine particles. In one method, Yamada [55] mixed UV-curable resins with ultra-fine Fe-based strong magnetic powders to give ink that could be printed on flexible films, fabrics, or paper to form electromagnetic shields. Polyester acrylate, epoxy acrylate, or urethane acrylate resins were used as the UV-curable binder and the ultra-fine magnetic powder was mixed at 80–100 vol % (based on the binder resins) [55]. In a second method, Suwabe et al. [56] prepared aqueous ink-jet inks, with good anticlogging ability and smudge prevention, by mixing aqueous dispersion of non-film-forming ultra-fine inorganic or synthetic polymer particles (e.g., PMMA particles) with pigments and film-forming resin fine particles. In a third method, Sawada et al. [57] and Uraki et al. [58] prepared ink-jet aqueous dispersion inks containing dispersed colored resin particles with an average diameter of 50–300 nm that were prepared by kneading organic pigments with water-soluble inorganic salts and water-soluble solvents in water and mixing with aqueous dispersions containing fine resin particles. The inks were easily filtered through a 0.45-μm membrane to form ink showing good discharge ability and transparency.

Composite structures consisting of metallic nanoparticles coated with organic polymers or organic polymer blend nanoparticles have been reported [59–67]. Funaki et al. [59] prepared a metal-organic polymer composite (especially porous) structure composed of a microphase-separated structure from a block copolymer in which a metalphilic polymer chain and a metalphobic polymer chain are bonded together at each end, and ultra-fine metal particles (<10 nm) were contained in the metalphilic polymer phase of the microphase-separated structure. Preferred polymers are a poly(2-vinylpyridine) and 2-vinylpyridine-isoprene block copolymer. The composite structures just mentioned are useful as functional material (e.g., catalyst) in heterogeneous catalysis. Ehrat and Watriner [60] prepared thermoplastic polyolefin or olefin copolymer powders with average particle size of 80 to 120 μm, by grinding in an impact mill together with fillers, such as Al, Mg and/or Ca hydroxides, carbonates or oxides. The composite mixtures are useful as highly filled molding compositions for battery electrodes or as powder coatings. Tamura [61] developed anisotropic magnetic-permeable composites. The composite contains ultra-fine particles of ferromagnetic Fe oxide that are smaller than single domain sizes dispersed in a solid organic polymer as oriented in domain direction of the particles and substantially separated from each other. The composites are prepared by dispersing the particles in a monomer and polymerizing the monomer in a magnetic field. Far-infrared radiation-emitting bodies from polymer microparticles and inorganic compounds have been reported [62]. The bodies are prepared from polymer particles with ultra-fine inorganic particles (e.g., Al_2O_3 or SiO_2) bonded to their surfaces. The bodies are useful for accelerating fermentation, preserving fresh food, and promoting plant growth. A typical method for producing the radiation-emitting bodies involves mixing an aqueous dispersion of PVC particles (2 μm) with $AlCl_3$ and NH_4OH to give PVC particles with adhering alumina hydrate particles (0.01 μm). The product can be extruded to give a film that is capable of far-IR radiation emission. The preparation of composite ultra-fine organic polymer mixture particles has been reported by Kagawa [63]. The composite particles were prepared by dissolving two different organic polymers in a solvent with a boiling point higher than the melting point of the polymers, and collision crushing with pressure to give particles with diameter <0.05 μm. The composite particles have good film-forming property (e.g., formed 5–10 μm film on Al foil that can be readily peeled off from the foil). Ultra-fine polystyrene particles and their composites with other materials can be prepared by adding dropwise polystyrene (weight-average molecular weight 3,840,000) solution (approximately 0.0002% in C_6H_6) to the surface of H_2O and the solvent evaporated to give a thin layer of ultra-fine particles which could be collected by moving barriers. The particles are cumulated on a chrome plate at a surface pressure of 1 to 50 dyne/cm^2 to give composite materials having area occupied with the particles ranging from 11 to 90%.

Hayashi et al. [65] and Suda et al. [66] developed low-temperature, glass-coloring agents containing ultra-fine noble metal particle – polymer composite and ultra-fine colored polymer particles, respectively. The former coloring agents contain a composite of ultra-fine Au, Pt, Pd, Rh, or Ag particles dispersed in a polymer without coagulation, an organo-metallic compound for fixing the ultra-fine particles in glass, a printing binder, glass powder, and an organic solvent [65]. The low-coloring temperature decreases strain in the colored glass and improves its cutting property. The ultra-fine colored polymer particles developed by Suda et al. [66] are useful for electrostatic photographic image developing agents or cosmetics. The particles are prepared by mixing the pigments with COOH (or ester)-containing polyolefins and non-aqueous solvents and precipitating. Stirring the pigment-coated carbon black, Zn naphthenate, and saponified EVA polymer in organic solvents, evaporating, and precipitating, resulted in particles with a median diameter of 1.564 μm.

Super-paramagnetic composites have been developed by Tamura [22]. The super-paramagnetic composite material consists of ultra-fine particles of ferromagnetic Fe oxide smaller than sizes of single domain structures dispersed as substantially separated in an organic polymer, and prepared by dispersing the particles in a monomer and polymerizing the monomer. An aqueous dispersion of the ultra-fine particles is prepared by a chemical reaction, hydrophobic coating of the particles by adding a surfactant to the dispersion, separating the particles, and adding the mixture to the monomer. The resulting composite has an extremely low residual magnetization and coercive force [67].

Ultra-fine polymer particles can be prepared from vinyl polymers [68, 69], polyisocyanates and acrylates [70], polystyrene [71], and polyacrylates [72–74] by graft copolymerization [68], spreading and curing [70], dispersion polymerization [71], and by emulsion polymerization [69–74]. The resulting particles are useful for water-resistant coatings and films [68, 70, 72], adhesives [71, 73, 74], and for freeze-thaw cycle resistant films [69]. Ultra-fine particle polymer latex is obtained by emulsion polymerization using a redox-type polymerization initiator in the presence of a compound serving as polymerization inhibitor solution in the monomer [75]. Coagulation of rubber-modified polymer latexes has been reported by Kitayama et al. [76]. In [76], a 30% acrylonitrile-butadiene-Me methacrylate-styrene graft copolymer (melting temperature 90 °C) latex was coagulated with an aqueous solution of $CaCl_2$ (11 mmol/L) at a concentration of 12 mmol/L and 95 °C, separated, water washed, de-watered, and dried to form a powdered polymer without any particles ≤200 mesh. Ultra-fine, particulate polymer latex based on unsaturated monomers with an average size <100 nm, a cross-linked structure, and glass transition temperature lower than that calculated by weight fraction method can be used to give a film excellent transparency, smoothness, tack, water resistance, and mechanical strength [77]. The polymer particle properties are dependent on the surfactant used. The latex is useful as a component in paints, adhesives, binder, additive for hydraulic inorganic material, fiber processing, reinforcement for optical glass fiber, electroconductive film, paper making, and photosensitive compositions [77]. Stable, aqueous colorants, useful in cosmetics, writing inks, and textiles, are prepared by encapsulating ultra-fine, primary particles with polymers which are not substantially altered in the process [78].

Artificial stone compositions for high-gloss products resistant to chemicals, water, and weathering can be prepared by mixing: 1) hydraulic inorganic material, 2) SiO2-based fly ash with an average particle size 1–20 μm, 3) water-dispersible ultra-fine granular acrylic polymer with an average particle size 50–2000 nm, and 4) pigment [79]. The ultra-fine granular acrylic polymer was prepared by emulsion polymerization.

Ultra-fine particles can be dispersed in organic polymers to form composites that exhibit good transparency and stability, making the composites useful for selective wavelength-shielding optical filters and nonlinear optical materials [80, 81]. Yao and Hayashi [80] prepared ultra-fine particle-dispersing polymer compositions from the reaction (which forms ultra-fine particles) of metal compounds and chalcogenation agents in an organic solution of polymer bearing pyrrolidone groups and stabilizer, followed by removal of the solvent. A typical preparation of the ultra-fine particle-dispersing polymer compositions consists of injecting 0.5 ml H_2S (g) into a tube containing 3.08 mg $Cd(NO_3)\cdot2.4H_2O$ and 0.01 g polyvinylpyrrolidone in 5 ml MeOH. Displacing the H_2S with nitrogen gave a product containing dispersed CdS with size 60 angstroms, which showed absorbency at 510 nm. In another patent, Yao and Hayashi [81] reported cuprous halide-dispersed polymer compositions and manufacture of polymer compositions dispersed with ultra-fine cuprous halide powders to give a transparent film containing ultra-fine CuBr particles with average diameter of 8.5 nm.

Giannelis et al. [82] have published a comprehensive review covering recent references on polymer-silicate nanocomposites. Recently, considerable attention has been paid to this type of nanocomposite to afford model systems to study confined polymers or polymer brushes and because of various applications in technical and biomedical fields.

3
Gas Atomization Approach to Generation and Analysis of Polymer Powders

The production of micron-sized polymer powders from molten polymers is an attractive, facile, low energy, and economic process. Polymer powders with tailored characteristics such as particle shape, size distribution, and purity can be directly prepared from the molten state of polymers such as polyethylene-based waxes that can not be ground using conventional methods [10–12,46,83–84]. The Gas Atomization Process (GAP) for mass-producing high quality spherical

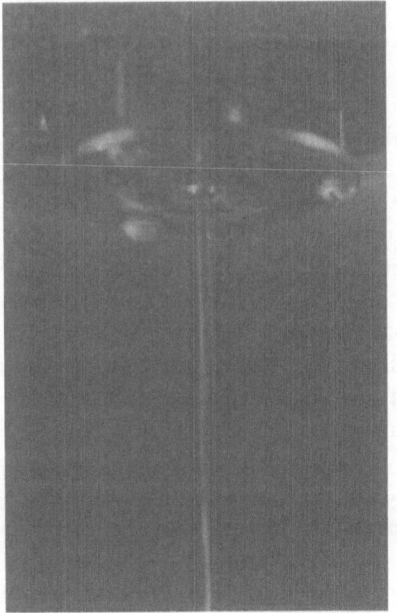 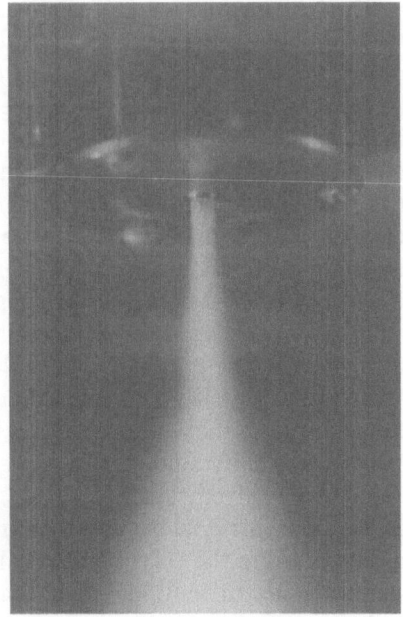

Fig. 1. The molten polymer stream before (*left*) and after (*right*) gas atomization

polymer powders involves using high pressure (approximately 7.6 MPa maximum) nitrogen gas and a specifically designed nozzle to atomize a molten stream of polymer into fine droplets that cool to form spherical powders [Fig. 1]. Powders with properties tailored to varying applications can be efficiently produced in short cycles by changing a few process control variables (described later) in a contamination-free environment, making the GAP a useful alternative to conventional grinding processes. These benefits of the process together with its flexibility, high throughput, and facile nature should make it highly attractive to industrial processes that must be capable of mass production, safe, and environmentally-benign operation.

The targeted applications of the powders include uses as powder spray coatings [2], formulating ingredients for functional coatings, and as raw materials for solid-state compacting of polymer alloys and composites [1, 85]. For these applications, the required properties of the powders include purity, uniform micron-sized particles with uniform size distribution, and spherical shape. These properties are essential for free-flowing powders with optimal surface area, leading to products with improved handling and performance capabilities.

Commercial organic polymer powders are produced by conventionally grinding extruded polymer pellets, often under cryogenic temperature conditions. Grinding is undesirable because it is expensive, highly energy-intensive, and susceptible to contamination from the grinding equipment and environmental pollution. Because of the erratic nature of the grinding process, it is almost impossible to control the quality and distribution of the powders and the size and shape of the particles.

The GAP method is an alternative route to mass producing polymer powders that eliminates most of the problems of conventional grinding operations. In addition, the simplicity and versatility of the GAP means that the equipment can be constructed from readily available construction materials such as steel (used in the crucible) and impact-resistant crystal-clear polycarbonate (used in the atomization chamber) (Fig. 2). The optical clarity of the latter allows direct real-time visualization of the atomization of the molten polymer as it exits the crucible. This process involves heating the material in a crucible until the desired atomization temperature is reached. Once the material reaches this temperature, it is forced out of the crucible through a circular channel (the pour tube) into the atomization nozzle, where it is atomized into fine particles by the high pressure nitrogen gas. The particles cool as they fall through the atomization chamber, forming micron-sized powders that collect in a vented chamber [Fig. 2]. Additional details of the GAP process has been reported [10–12, 46, 83–84].

GAP feasibility studies and process development efforts focused on using commercial polyethylenes (e.g., Hoechst Celanese's PE130 and PE520) [Ref. 86] as the model material because they are presently the largest volume commodity plastics used in the U.S. (over 9×10^9 kg are produced annually). The high consumption, low toxicity, low molecular weights (2000 to 10000 g/mol), and low melting temperatures (approximately 200 °C) of the PE130 and PE520 make them ideal materials for atomization. Thus far, only the low molecular weight

The Heating Zone 1
- Crucible and Components
- Temperature Control
- Pneumatic Flow Control

The Atomization Zone 2
- Atomization Nozzle
- Atomization Chamber

The Collection Zone 3
- Collection Chamber
- Exhaust Vent

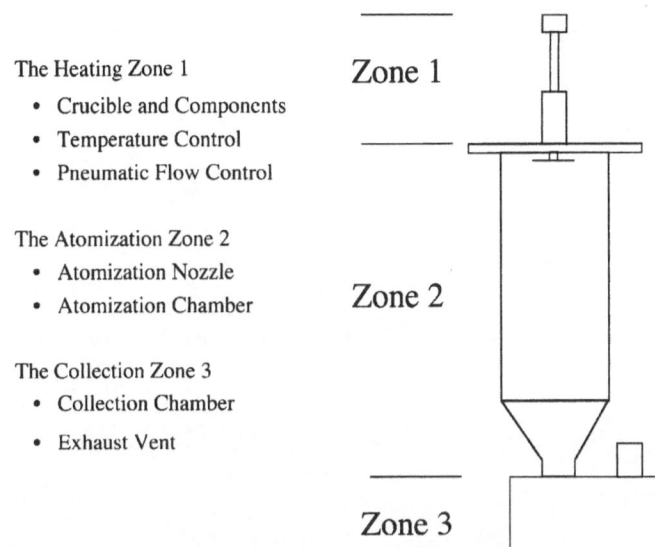

Fig. 2. Schematic of GAP showing the key zones and components

polyethylenes have been atomized into fine powders with changeable particle shapes and size distributions (0–250 μm). The studies conducted to date show that the quality and properties of the product powders depend on three key processing variables: 1) polymer melt temperature, 2) gas atomization pressure, and 3) melt stream size or pour tube diameter (Fig. 3). Unlike particles produced by conventional grinding, the particles in gas-atomized powders are spherical with smooth surfaces and near uniform sizes. Other particle shapes – such as whiskers and elongated spheres – can be produced under specific processing conditions such as using low atomization pressures (approximately 2 MPa). Typically, the whiskers have diameters of about 100 nm and lengths of a few millimeters.

To expand the GAP method to produce powders from other polymers with tailored powdered characteristics for wide applications, computer simulations of the process are needed. As mentioned, potential applications of the product powders include use as formulation ingredients for functional coatings tailored to specific biochemical engineering application areas, such as personal hygiene and beauty care products, packaging, and other disposable and/or recyclable plastic products. Other applications include polymer dispersions or emulsions in environmentally-friendly solvents, feedstock for solid-state compacting of polymer alloys, and powder spray coatings. As an example, the PE520 is designed for use in paints to increase matting effects and to mar resistance of the painted surface. Other advantages of using polymer powder additives in paints are improved sanding ability smoothness, and rheological properties; prevention of pigment settling and metal marking; and water repellence. Because of the

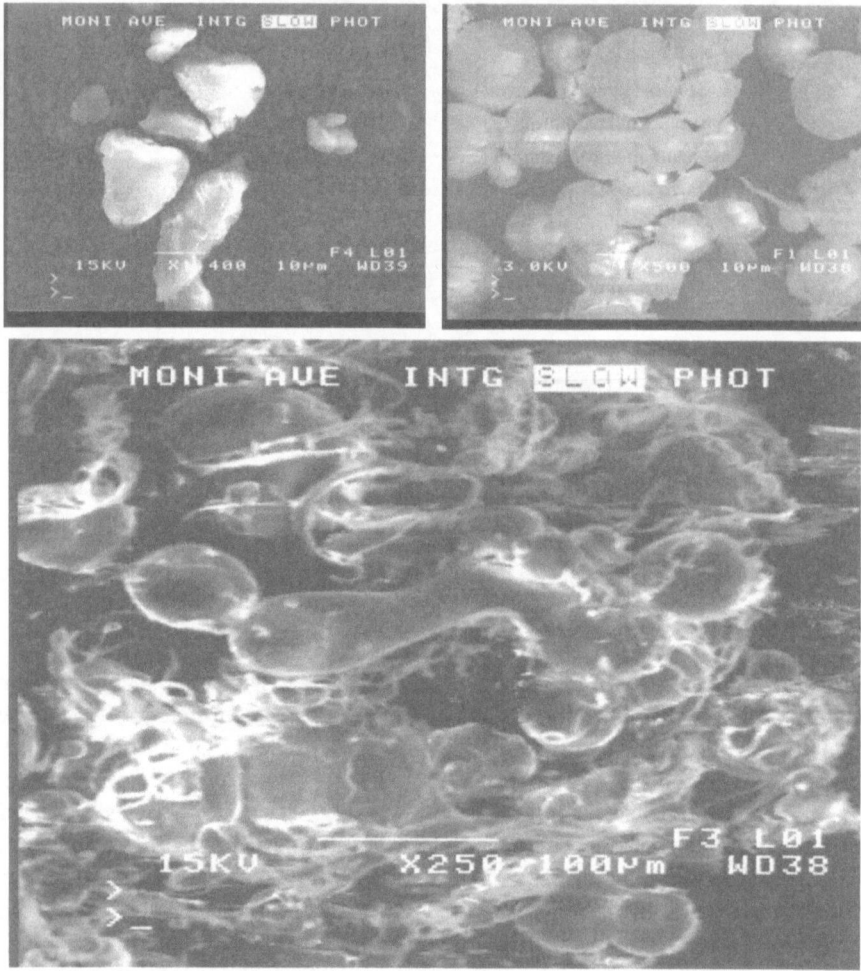

Fig. 3. SEMs of commercially ground PE-based wax (*left*), PE520 atomized at pressures of 7.6 MPa (*right*) and 2 MPa (*bottom*). (the PE520 was atomized at a temperature of 190 °C in both cases)

flexibility, versatility, and economy offered by GAP, it should be attractive to polymer manufacturers, processors, and end-users.

Because the polymers (PE130 and PE520) can be atomized in a relatively narrow temperature range (190–220 °C), temperature control in GAP must be precise to avoid potential thermal degradation of the molten polymer prior to atomization. This can be achieved by heating the polymer in the crucible under a blanket of nitrogen gas using precisely controlled band heaters with thermocouples strategically placed in the melt. Obviously, polymers that show different rheological properties under conditions that they are likely to encounter during atomization can be expected to atomize differently. At low pressures (approximately 2 MPa), for example, the shear induced by the gas jets on the molten polymer at the instant of melt disintegration is not enough to completely overcome the internal elastic stresses present in the molten polymer. This leads to the formation of whiskers and elongated spheres rather than absolute spheres, (Fig. 3). For the polyethylenes studied thus far, it has been found that the formation of whiskers and elongated spheres can be avoided by using high gas atomization pressures (approximately 7.6 MPa), (Fig. 3). It appears that a mixture of whiskers and spheres would be ideal for making self-reinforced polymer powders that can be used for applications requiring improved mechanical properties. Investigations on expanding the use of GAP to other polymers with vastly different thermal and rheological properties are needed.

More recently, 50/50 blends of PE130 and ultra-low melting phosphate glass composition have been successfully atomized under conditions that were used to atomize the pure PE130 polymer. This result confirms the expectation of the broad application of GAP to many fields, such as producing polymer alloys, glass-polymer alloys, in situ composites, and related products with tailored properties. These products could be used in many areas, such as decorative or protective coatings, polymer-supported heterogeneous catalysts, and in producing lightweight structural composites. The structural composites can be easily fabricated by applying established solid-state powder compaction methods to the gas-atomized composite powders to form compacts with varying shapes and sizes.

The research conducted thus far has provided valuable insight into GAP diagnostic control systems, dynamics, and mechanisms of powder formation. This knowledge can be used to expand the method to include the production of other kinds of materials with desirable properties for beneficial uses. The desirable properties of the powders include the following: purity, particle shape, particle size, and size distribution. The many teething problems of GAP when it is applied to polymers and composites are now understood and can be controlled and managed in order to produce powders with tailored characteristics. The technology of GAP has now advanced to a stage of finding more applications in the areas of polymer engineering, composite engineering, and of scaling up to mass production of the fine polymer powders.

3.1
Experimental Results and Analysis

In this section, we will answer this question: Why do melt stream size and temperature affect the atomization results? In addition, one might ask the following questions: 1) Why do materials atomize differently? and 2) What is the key material property responsible for the difference? Furthermore, we will speculate on the broad application of the GAP process to higher molecular weight polymers than the ones atomized thus far.

The initial melt stream size used was the 4.763-mm pour tube. This pour tube provided the process dynamics that created a density gradient within the stream as evident in the high-speed video of the GAP process [87]. This density gradient could possibly be attributed to the formation of spheres on one side of the stream while the other side produced fibers. By using smaller pour tube sizes, we hoped to minimize (possibly eliminate) this density gradient. The 3.175-mm pour tube worked well in eliminating production of many of the fibers, but the 1.588-mm pour tube produced even more fibers. The reason that the 1.588-mm pour tube produced numerous fibers is that the melt stream size was too small to flow out in an even, radial pattern into the pathway of the atomizing gas jets. Therefore, there were thin areas within the melt stream that did not carry enough mass and/or sufficient stress to break-up into spheres when contacted by the sonic gas, leading to the formation of the microfibers. Based on this observation, experiments with modified pour tubes to maximize the radial flow of the polymer melt are needed.

The small temperature window in which the material can be atomized greatly limits the operating range of GAP. The material must be heated to a temperature high enough to avoid freeze-up within the pour tube but below its onset temperature for thermal degradation ($T_{d, onset}$). Although a significant amount of the melt is atomized when it comes in contact with the sonic gas flow, further disintegration and refinement can occur along the trailing shear boundary layer created by the sonic gas. Because the molten polymer can be atomized within a small temperature range, much of the material freezes upon initial contact with the atomization gas at −90 °C. Thus, the shock-enhanced disintegration of the polymer melt [88] is not able to refine most of the polymer powders. This finding is supported by the high-speed videotape of the atomization process, which shows that the polymer melt disintegrates and freezes instantly when it comes in contact with the atomization gas. It is believed that this fast quenching of the polymer melt contributes to the reduction in the observed percent crystallinity of the atomized samples. Note that the amount of particle refinement increases with increasing temperature up to a limit dictated by the $T_{d, onset}$. At higher temperatures, more material remains in the molten state after initial contact with the atomization gas and, therefore, secondary disintegration and refinement can occur. This expectation explains why the most favorable atomization results for PE130 and PE520 occurred at temperatures near the $T_{d, onset}$ for each material.

The physical properties of PE130 and PE520 [11,84,86] and the different atomization results already described indicate that the higher density of the more crystalline PE130 is a key material variable that favors better atomization of PE130 over PE520. The density of the Newtonian PE-based polymers atomized thus far, was found to effect the polymer melt break-up dynamics during atomization [11, 84]. A denser material is more likely to remain together when it disintegrates into the lowest free surface energy structure of a sphere. A less dense material would be more likely to be drawn into fibers. The waxy nature of PE520 makes it impossible to grind, causing it to be freely drawn into fibers during atomization under certain conditions. The lower viscosity and percent crystallinity of the two PE-based polymers after atomization suggests molecular chain scission and reorganization caused by the action of the sonic atomization gas jets on the molten polymer at the instant of melt disintegration. However, a definitive statement cannot be made at this time. This suggestion is consistent with the fast quenching of the molten polymer caused by the –90 °C atomization gas. Preliminary molecular dynamics simulations of flow of PE-based melts through various channels appear to confirm the expectation of molecular chain scission and/or reorganization under certain imposed conditions [89].

Clearly, polymers that show different rheological properties under conditions that they are likely to encounter during atomization can be expected to atomize differently. For example, at low pressures (2 MPa or 300 psi), the shear induced by the gas jets on the molten PE520 at the instant of melt disintegration was not enough to completely overcome the internal elastic stresses present in the polymer. This led to the formation of whiskers and elongated spheres rather than absolute spheres (Fig. 3, bottom) [12]. Studies of the influence of melt viscoelasticity of high-molecular weight polymers on their atomization results is a matter for future investigations.

Although we have not yet atomized other high-viscosity, high-molecular weight polymers, it is interesting to speculate on the application of GAP to these polymers. Because of the high viscosity and shear-thinning or pseudoplastic nature of high molecular weight polymers, formation of microfibers (with very high aspect ratios) from the molten polymer atomization is expected to be favored over formation of powders from these specific polymers. Based on current GAP research, the indications are that high-viscosity polymer melts can be expected to be suitable for making the microfibers using essentially the same approach but with some modification to the polymer atomization equipment. The modification is essential to accommodate the nonlinear viscoelastic behavior of high molecular weight polymer melts under temperature and shear deformation conditions that they must encounter during polymer atomization.

The force (or more correctly, stress) experienced by the molten polymer during atomization is a function of its elongational (tensile) viscosity, volumetric flow rate, and velocities of the atomized microfibers like others have reported for the spunbonding process [90]. The spunbonding process involves extruding molten polymers through spinnerets, into an aspirator, and onto a conveyor belt equipped with a vacuum [91,92]. Unlike the GAP process, the aspirator in the spunbonding process is used primarily for the fiber draw down. The high pressure and the high sonic velocity of the atomizing gas in GAP will allow formation of highly oriented microfibers with very high aspect ratios. This expectation appears to be consistent with the low-pressure (2 MPa) atomization results on PE520 already described. The tendency to form polymer microfibers in the GAP is supported by the theoretical expectation that for crystalline polymers, the orientation of the crystallographic axes (a, b, and c) and the crystalline content is controlled by the stress experienced by the polymer melt at the instant that the atomizing gas contacts the molten stream of polymer [93]. This theoretical expectation is consistent with the observed changes in the percent crystallinity of the PE-based polymers (described in this paper) before and after atomization and by real-time videography of the polymer atomization process. Therefore, with the proper choice of polymer atomization processing conditions – type of nozzle and its alignment relative to the orientation of the gas jets, melt temperature, atomization pressure, melt stream size or pour tube inside diameter, degree of crystallinity of the polymer, etc. – the GAP can be used to make powders with tailored powdered characteristics (such as spheres and microfibers) with ultra-high aspect ratios (not possible now with available methods) for many beneficial uses.

3.2
Neural Network Modeling of GAP

This section discusses computer simulations of the GAP to expand the method to produce powders from other polymers with tailored powdered characteristics for wide applications. To allow both forward and reverse predictions of the weight fractions of the polymer particles having the specified properties such as size ranges and shapes, we report the application of partial least squares regression and computational neural network methods to analyze the experimental data [10, 11, 84]. The ultimate goal is to train the polymer atomization technique using k-fold and a Levenberg-Marquardt optimization method such that the shape and size distribution of the product particles can be predicted within a maximum absolute percent error estimated from the above analysis performed on real experimental data. The predictive models (described later) are expected to be useful in expanding the GAP technology to a wide range of polymers and applications.

Computational neural networks (CNNs) are model-free estimators with an exceptional ability to perform multi-dimensional, non-linear vector mappings [94–100]. They are robust against noise and tend to be immune to violations of assumptions that would cripple many traditional methods. These qualities make CNNs general purpose tools for data analysis- modeling, diagnosing, or controlling complex processes such as GAP. For example, CNNs offer powerful complementary modeling capabilities for molecular simulation by either: 1) replacing some part of another modeling capability, or 2) emulating the entire modeling technique. Successful implementations have been reported for most molecular modeling tools (molecular dynamics, normal mode analysis, Monte Carlo, quantum chemistry, etc; see Fig. 4.) with notable improvements in computational efficiency and generality. A recent review discusses these applications with regard to chemical [94] and materials science [95]. Below we give a brief introduction of the CNNs most commonly used in computational chemistry applications.

In the area of process systems engineering there has been a rapid increase in academic and industrial interest in CNNs. CNNs have been used throughout the chemical-processing and related industries for assisting or solving problems related to the production of polymers [101], steel [102, 103], ceramics [104], sem-

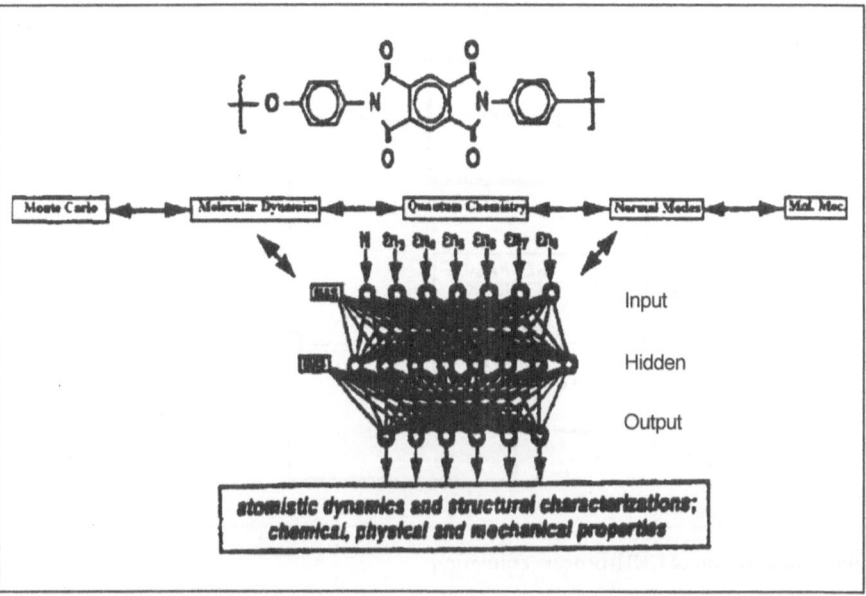

Fig. 4. Schematic illustrating a synergistic integration of CNNs with modeling and simulation tools

iconductors [105], and composites [106]. Applications include reaction and chemical process monitoring, optimization, diagnosis, control, chemical composition analysis, pattern recognition for predicting properties of materials, characterization and modeling of chemical reactor processes, product design, and the classification of multi-component feed stocks [107–113]. The literature on this subject is quite extensive (see for example Refs. 109–113). Over 800 articles have been published since 1986 and hundreds more are published each year, which at least implicitly demonstrates the success of CNNs for these types of tasks. The suitability of CNNs for such problems stems from the ability to model complex and nonlinear relationships that might exist between process data and control parameters.

Often, a manufacturing technology involves highly non-linear fabrication processes that make experimental data expensive to obtain. Since it is important to know as much as possible about the behavior of a process, a considerable amount of research has been devoted to the theory of process control. Various types of statistical analyses, pattern recognition, CNNs, knowledge-based systems, and linear models have been developed and applied to the problem of system modeling, optimization, fault diagnosis, and control. In particular, employing CNNs to model complicated processes has been very successful. In this approach, sometimes called system identification, a CNN is used to emulate the process (see Fig. 5) and/or its inverse [114–117]. System identification requires the investigator to first perform a series of statistically designed characterization

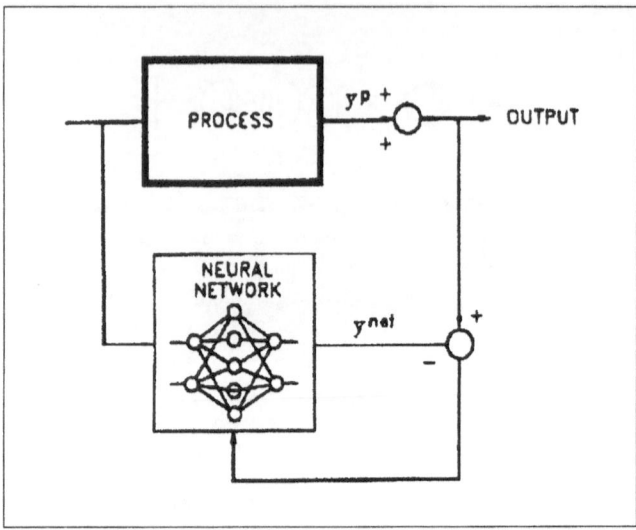

Fig. 5. Illustration of CNN-process emulation

experiments (some type of factorial experiment) in order to collect data suitable for training a CNN. The goal of system identification is to either provide a model that allows simple and cheap investigation of the sensitivities of the input variables (optimization) or to use the model in a feedback loop for controlling the process. A natural extension of CNN assisted modeling of a process is to use this model as a means for optimizing the process [118]. By using the functional relationship between controllable input parameters and process responses, procedures for better performance can be determined.

3.2.1
Neural Network Fundamentals and Methods

By storing information in a weighted distribution of connections, a typical CNN uses available data to "learn" the essential relationships between given inputs and outputs. A learning algorithm provides the rule or dynamical equation that changes the distribution of the weight space to propagate the learning process. CNNs have several essential constructs that define their operation: nodes (simple processing units), transfer functions (generally nonlinear and bounded functions), connection weights, and a learning algorithm. One possible arrangement of the nodes is an architecture that describes a multilayer feedforward network: a set of nodes placed into two or more layers. There is an input layer and an output layer, each consisting of at least one node. The nodes in the input layer do not perform any actual processing but serve only to distribute the input to the next layer. There is usually one or more hidden layers (layers of nodes between the input and output). The term feedforward means that the inputs to the nodes in each layer come exclusively from the outputs of nodes in the previous layer, and the outputs from these nodes pass to nodes in the following layer. Each node in the network has a number of weighted connections to other individual nodes and an input signal is propagated through the system until it emerges as a network output. An optimization procedure that adjusts the weights connecting the nodes in order to minimize the difference between the output and the target (the desired result) is called the learning algorithm (see Fig. 6). Backpropagation was the first practical method for training multilayer feedforward networks and is still the most popular learning algorithm [99, 100].

The backpropagation algorithm is an example of supervised learning – training with a teacher, that is, with known answers for representative examples. This algorithm adjusts the weights based on a gradient descent minimization of an error function (usually the sum of squared errors). The goal is to "teach" the network to associate specific output to each of several inputs. Having learned the fundamental relationship(s) between inputs and outputs, the neural network should then be able to produce reasonable output for unknown input (called generalization). The steps in the standard on-line backpropagation algorithm are:

1. Initialize all node connection weights w_{ij} to some small random values.
2. Input a training example V_i^m and corresponding output values V_i^T, where m is the layer number, i is the node number, and T represents the target or desired output state.
3. Propagate the initial signal forward through the network using:

$$NET_j^m = \sum w_{ij}^m V_i^{m-1} + \beta_j \quad \text{and} \tag{1}$$

$$V_j^m = F\left(NET_j^m\right), \tag{2}$$

where w_{ij} is the connection weights between nodes i and j, V_i^{m-1} is the signal from node i and layer m-1, β is the threshold or bias value of the node, and F is a transfer function (a function that, when applied to the input of a node, determines its output), usually taken as a sigmoid function, most commonly the logistic function:

$$F\left(NET_j^m\right) = 1 / \left[1 + \exp\left(-NET_j^m\right)\right]. \tag{3}$$

This function is continuous and varies monotonically from a lower bound of 0 to an upper bound of 1 and has a continuous derivative. The transfer function in the output layer can be different from than that used in the rest of the network. Often, it is linear, f(NET)=NET, since this speeds up the training process. On the other hand, a sigmoid function has a high level of noise immunity, a feature that can be very useful. Currently, the majority of current CNNs use a nonlinear transfer function such as a sigmoid since it provides a number of advantages. In theory, however, any nonpolynomial function that is bounded and differentiable (at least piecewise) can be used as a transfer

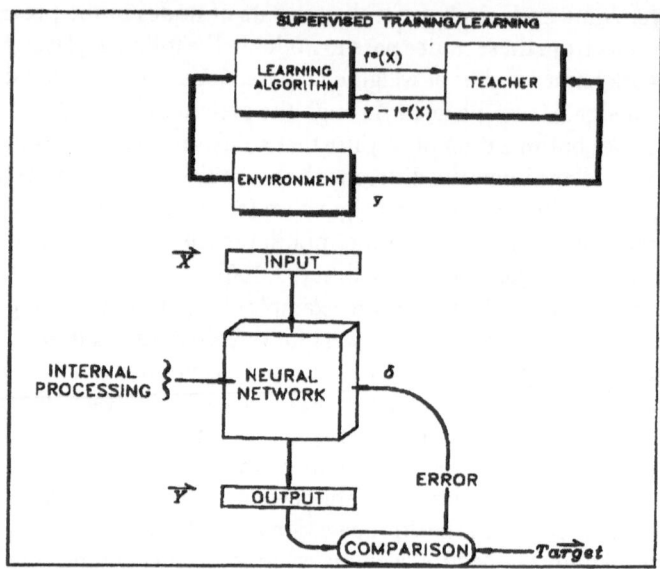

Fig. 6. Schematic illustrating the basic concepts for the CNNs supervised learning algorithm

function without altering the CNN capabilities of universal approximation [119]. Other locally bounded and differentiable functions, such as the hyperbolic tangent and a scaled arctangent have also been extensively used and numerous others have been suggested. In most cases, it has been found that the exact shape of the function has had little effect on the ultimate power of a CNN, though it can have significant impact on the training speed.

The feedforward propagation (Eq. 2) is continued for each i and m until the final outputs V_i^o have all been calculated.

4. Compute the deltas (δ) for the output layer, defined as:

$$\delta_i^o = -\partial E / \partial NET_i^o = -\left(\partial E / \partial V_i^o\right)\left(\partial V_i^o / \partial NET_i^o\right) = F'\left(NET_i^o\right)\left[V_i^T - V_i^o\right], \qquad (4)$$

where the error function $E = \sum \left(V_i^T - V_i^o\right)^2$ and $F'\left(NET_i^o\right)$ is the derivative of the transfer function with respect to the activation NET_i. For the logistic function the derivative is:

$$F'\left(NET_i^o\right) = \partial F\left(NET_i^o\right) / \partial NET_i^o = F\left(NET_i^o\right) \cdot \left[1 - F\left(NET_i^o\right)\right], \qquad (5)$$

where $F\left(NET_i^o\right)$ is given by Eq. (3).

5. Compute the deltas for the preceding layers by propagating the errors backward:

$$\delta_i^{m-1} = F'\left(NET_i^{m-1}\right)\left[\sum w_{ji}^m \delta_j^m\right], \tag{6}$$

for all m=m, m-1, m-2, ..., until it has been calculated for each layer.

6. Using:

$$\Delta w_{ij}^m = \eta \delta_i^m V_j^{m-1}, \tag{7}$$

update the connection weights to:

$$w_{ij}^{new} = w_{ij}^{old} + \Delta w_{ij}. \tag{8}$$

7. Return to step (2) and repeat for another input example.

This process (steps 1–7) is continued until the network output satisfies some ending criteria. In practice this type of iterative approach can take numerous epochs (cycles through the whole data set) before a reasonable error is reached. This is one of the disadvantages of the standard backpropagation algorithm. Fortunately, there are a number of methods that can help alleviate this pathology. For example, we have had very good success with the Levenberg-Marquardt compromise to Newton's method for optimization problems. This method basically interpolates between gradient descent and Gauss-Newton methods depending on the distance away from the minimum. Gradient descent is used for minimizing the error for positions that are far away from a minimum and Gauss-Newton is used for those close to a minimum. The approach is quite practical for <100 variables since it is well-known that second order or quasi-second order methods converge much faster than gradient descent close to a minimum (the error surface is quadratic – a well-defined Hessian matrix) but are slower when the error surface is not parabolic (far away from the minimum – the Hessian is not positive definite). However, if the number of weights for the optimal neural network architecture is greater than 100, the Levenberg-Maquardt method becomes very inefficient [120]. In this case, we have found that the Polak-Ribiére conjugate gradient method [121] with Powell restarts [122] provides an excellent optimization method for neural networks [123]. Adding a stochastic perturbation term to this method, much in the same spirit as Langevin dynamics [124], can also give improved performance (helps to escape from local minima in the hope of finding a global minima).

3.2.2
Designing Neural Network Solutions

Most CNNs perform a similar task: an input to output vector mapping. The primary difference is only in the way this task is performed. Typically, only one or a small number of paradigms are needed to treat most practical problems. Multilayer, feedforward CNNs trained with the backpropagation algorithm are by far the most commonly applied. This is primarily due to the fact that these networks can approximate any Borel measurable function (they are universal function approximators) [125], have a proven history, and are easy to implement. In the following section, we discuss maximizing the probability of success when using backpropagation CNNs to solve practical problems.

3.2.3
Data Pre-Processing

The performance of a CNN is strongly dependent upon the patterns used to train it (training set). If the network is to meet expectations, the training set must provide a full and accurate representation of the problem domain. Several goals should be satisfied: 1) every class must be represented; 2) within each class, statistical variation must be adequately represented; and 3) input vectors should have a small number of components, if possible. If advanced knowledge on the behavior of the data is available – for example, if some patterns are easier to classify than others-, shaping the data by weighting patterns (over representation) can often be advantageous. If a sufficient amount of numerical data is available, these guidelines provide a reasonable way to collect a training set. In addition, there are a number of methods which can assist in the task of selecting an optimal training set, such as combined optimization techniques [126], Kohonen self-organizing maps [127], the ΔISB method described by Plutoswki & White [128], and statistical resampling techniques [129].

Once a suitable training set has been collected, preprocessing the input data often plays a critical role in the CNN's ability to carry out the desired task [130–132]. Scaling or normalizing the data can help improve the performance by removing insignificant characteristics. In addition, scaling the data to values that fall in the range of the bounds of the transfer function is generally recommended, especially when using output nodes with a bounded transfer function. Correct scaling of input data can be enormously useful for enhancing network performance. However, if done incorrectly, it can render a data set meaningless.

The mean and standard deviation generally are not part of the important aspects of the data and may obscure the issue and complicate the networks' task. A good method for removing both is the Z-score procedure, which involves subtracting the mean and dividing by the standard deviation. This removes all effects of offset and measurement scale. A simple linear mapping can be used to normalize the Z-scores to the bounds of the transfer function.

Supervised neural networks require a training and testing data set. Unbiased selection of these sets from the available data can be achieved by using the bootstrap resampling method [129]. This method, founded in statistics, is a powerful procedure for determining the best estimator for small data sets. Bootstrap resampling is basically random sampling with replacement-a data set of N examples is randomly sampled N times to create a new data set with N examples. The new data set will have the possibility of sample repetition and a test set can be generated by comparing the new data set with the original one and selecting those examples that are unique. Thus, the bootstrap method can be used to produce a number of different training and test sets and the error estimate is taken as the average performance on the data set ensembles. In general, it has been found that about 200 iterations of the bootstrap estimates are needed to obtain a good representation.

Bootstrap resampling is only one of several methods that provide desirable properties in error rate estimates. Another method that works very well for neural network training and testing is the jackknife method. The general procedure is to take one (it also can be generalized to k samples) sample out of the available data as a test set and train on the remaining ones. This is repeated for all of the samples, producing an ensemble that is the same size as the original data set. This procedure can be used like the bootstrap but it requires fewer runs.

By using statistical resampling (bootstrap or jackknife), a CNN can be trained on an ensemble of different training sets and the performance (cross-validation) evaluated on the complementary testing sets to determine a measure of the true performance (the average over the ensemble). Some added benefits on error performance can also be obtained by averaging the network's error over an ensemble of initial connection weights (generally, one set of connection weights will lead to convergence to a different local minima than another set). Called the ensemble average method, this approach not only gives a better estimate of the true error (unbiased) but also reduces the size of the neural network (number of hidden nodes) over that required for any single training/testing computation [133]. It also tends to smooth) the effects of over-fitting (variance reduction) by averaging in function space instead of parameter space. This procedure is obviously more computationally demanding than one consisting of a single run but adequately makes up for this pathology since it provides an optimal and unbiased representation of the performance capabilities of a CNN.

3.2.4
The Initial State

An area of importance for enhancing the convergence of the backpropagation algorithm is the initial selection of the weights [134]. Backpropagation is very sensitive to initial conditions. If the weights are located within an attraction basin of a local minima attractor, convergence can be rapid; if the weights start in a flat region of the error surface, it can be slow. In general, the weights W are initialized to small zero-mean random values ($W \in [-r,r]$, where r <1). The reason for starting from small random weights is that large weights tend to prematurely saturate nodes in a network and cause the backpropagation algorithm to get stuck (network paralysis). Initial selection of random weights also introduces a symmetry breaking mechanism that prevents nodes from adopting similar functions and becoming redundant. It has been found that premature saturation can be avoided by initializing a network on a node by node basis. This technique chooses weights randomly and uniformly distributed on the interval $W \in (-3/\sqrt{f_i}, 3/\sqrt{f_i})$, where f_i is the number of inputs (fan-in) for node i.

3.2.5
Successful Training Techniques

One of the dangers in backpropagation training of CNNs is the tendency to over-fit the training set. Over-fitting is due to the combination of the nonlinear modeling properties of the network over long training times. It can also result from a training set that does not totally represent the relevant population or an oversized network Cross-validation can minimize over-fitting. A cross-validation data set is drawn from the same population as the training set, but it is not used for training. In order to avoid over-fitting, the network is trained using the training set but the error is monitored on the cross-validation set. When the error of the cross-validation set reaches its minimum, training stops. Combined with complexity regulation, an optimal generalization error can be obtained. However, since the network is never trained on the cross-validation set, it may be missing valuable information about the distribution of the data. In addition, extreme care must be taken not to introduce accidental bias by the choice of division used to create the cross-validation and training data. The best method that we have found is the ensemble average method, described in the section on data preparation and representation.

We have not yet addressed the important problem of determining or selecting the number of layers (architecture) and hidden nodes (topology) for a feedforward CNN. Some useful guidelines based on theoretical and empirical results can be given. First, always start with only one hidden layer [135]. There are a few rare situations in which two hidden layers may be preferable, but more than two are never theoretically needed [136]. Thus it is relatively simple to determine the best architecture. Selecting the number of hidden layer nodes for a feedforward neural network is best done by the brute force method. Most empirical rules are application specific and can often lead to extremely poor guesses (although the geometric pyramid rule, for n inputs and m outputs use a hidden layer of $\sqrt{(nm)}$ or for two layers, $m(n/m)^3$ and $m(n/m)^{3/2}$ can often be instructive). Common wisdom suggests that a network should not have more free parameters (connections) than training examples. Three to 10 examples per connection is often recommended. A reasonable and empirically proven method for finding the optimal number of hidden nodes is to start with a few nodes (one, for example) and incrementally add nodes until good performance is achieved (a bottom-up approach). In general, most applications never require more than 15 hidden nodes; those applications that do require more nodes can generally benefit from breaking the problem into simpler pieces (divide and conquer approach).

3.2.6
Other Neural Network Paradigms

Feedforward CNNs trained with the on-line backpropagation algorithm or one of its many variants is the most common method used to solve or treat scientific and technological problems. Although no other CNN paradigms have demonstrated the time-tested range of applications of backpropagation, there are a number that do provide potentially useful features for general purpose problem solving. Another approach to training feedforward neural networks can be viewed as a method for general curve-fitting as opposed to the stochastic approximation achieved by on-line backpropagation. The idea is to use a linear combination of functions to find a surface in multi-dimensional space that provides a good fit to some set of data. Typically this approach is cast onto a neural network representation by using a network of functions $\ddot{Y}$, the so-called basis function neural networks (most commonly the radial basis function).

The radial basis function (RBF) network is a two layer network whose output nodes form a linear combination of non-linear basis functions computed by the hidden layer nodes [137–143]. The basis functions in the hidden layer produce a significant nonzero response only when the input falls within a small localized region of the input space (receptive field). In general, the hidden layer nodes use Gaussian response functions, with the position (w) and width (σ) used as variables:

$$f_i(x) = \exp\left[-(x-w_i)^T(x-w_i)/2\sigma_i^2\right], \tag{9}$$

where superscript T signifies the transpose, w_i are weights, and σ_i is the width parameter of the Gaussian (standard deviation of the response curve of the node). The output layer nodes compute a weighted sum of the results from Eq. (9),

$$\sum f_i(x)W_i , \tag{10}$$

and can be optimized using supervised training (generally a gradient descent or the Least Mean Squares algorithm). The parameters for the Gaussian nodes are generally adjusted using an unsupervised training method, self-organizing maps or other clustering methods such as K-means. The overall training algorithm generally involves adjusting the values of w, σ, and W using a combination of supervised and unsupervised training.

Radial basis functions networks are good function approximation and classification as backpropagation networks but require much less time to train and don't have as critical local minima or connection weight freezing (sometimes called network paralysis) problems. Radial basis function CNNs are also known to be universal approximators and provide a convenient measure of the reliability and confidence of its output (based on the density of training data). In addition, the functional equivalence of these networks with fuzzy inference systems have shown that the membership functions within a rule are equivalent to Gaussian functions with the same variance (σ^2) and the number of receptive field nodes is equivalent to the number of fuzzy if-then rules.

Another basis function network that has gained considerable popularity, called the generalized regression neural network (GRNN), has foundations in non-linear regression theory [144]. This network provides a method for estimating f(x,y) from information in the training set by deriving a probability density function (using parzen windows [145]) with no preconceived idea about its form. This general technique can even estimate functions that are composed of multiple disjoint non-Gaussian regions in any number of dimensions The architecture and output of this network is identical to that of the radial basis function network, except that it has a normalization term, $\Sigma f_i(x) W_i / \Sigma f_i(x)$ which replaces Eq. (10). The primary difference is the method for determining the weights, W, and Gaussian centers, w. GRNN's train in one pass by assigning the centers of the Gaussian functions w to the training vectors and the weights W_i to the value of the target for input i and component j. In practice, however, the value of σ in Eq. (9) must be optimized. This is easily achieved using a brute force method (selecting a fine grid of values for sigma and using the best one) or by implementing a gradient descent algorithm [146].

3.3
Discussion of Neural Network Results of GAP

Three input variables- X1 (temperature), X2 (melt stream size), and X3 (material type)- were changed in a systematic manner and five output variables (Y1 through Y5) based on weight fractions of powders having the micrometer ranges were measured as described previously [11–84]. The results are shown in Table 2. In this table, T = atomization temperature, S = stream size, Y1= weight percent for 0–53 μm, Y2= weight percent for 53 μm to 106 μm, Y3= weight percent for 106 μm to 150 μm, Y4= weight percent for 150 μm to 295 μm, Y5= weight percent for 295 μm to 600 μm.

The experimental data are shown in Table 2. To methods were used to determine predictive models for the weight fractions of the powders having micrometer ranges specified by Y1 through Y5 (forward prediction mode) and for prediction of process variables such as X1 (temperature), X2 (melt stream size), and X3 (material type) based on weight fractions of powders having the micrometer ranges already specified above (reverse prediction). In the forward prediction mode, two methods were used to develop a model, partial least squares regression (PLS) and CNNs (three inputs, five hidden, and five outputs trained using k-fold cross validation and a Levenberg-Marquardt optimization method) as already described. The results obtained are shown in Table 3. In this table, the errors are reported as the average absolute percent error (compared to experiment) and the maximum absolute percent error over an ensemble of five trials consisting of 40 training and 10 examples for training and testing, respectively (chosen randomly: the K-fold cross validation or leave-k-out procedure). In the reverse prediction of X1 through X3 based on Y1 through Y5 experimental data, the two methods were used to develop the model: PLS and CNNs (five inputs, five hidden, and three outputs; trained using the cross validation and optimization method already described above. The results obtained are also shown in Table 4. The errors are reported as already described.

Based on the neural network model of the GAP process, both forward and reverse predictions of conditions for maximizing the weight fraction of the pow-

Table 2. Experimental data

Run #	X1	X2	X3	Y1	Y2	Y3	Y4	Y5
1	180	0.063	130	9.2	13.6	9.2	13.9	54.1
2	195	0.063	130	10	13.9	9.8	13.9	52.3
3	200	0.063	130	8.1	10.7	8.6	14.7	57.9
4	220	0.063	130	8.2	12.6	9.1	14	56.1
5	170	0.125	130	6.5	18	17.5	38.9	19.1
6	175	0.125	130	9.9	20.8	17.9	32	19.4
7	180	0.125	130	6.9	21.5	18.2	37.7	15.7
8	185	0.125	130	9.1	23.1	18.2	32.6	16.6
9	190	0.125	130	11.7	27.7	21.9	27.3	11.4
10	195	0.125	130	11.2	31.8	17.9	22.6	16.5
11	205	0.125	130	13.3	25.1	18.3	27	16.3
12	210	0.125	130	10.5	28.8	16.7	31.3	12.7
13	215	0.125	130	11.3	26.1	19.5	29.6	13.5
14	220	0.125	130	10.3	20.7	14.7	32.5	21.8
15	170	0.188	130	4.4	12.1	13.2	41.1	29.2
16	175	0.188	130	5.3	11.4	14.5	40.6	28.2
17	180	0.188	130	6	15.5	15.2	37.9	25.4
18	185	0.188	130	6.3	18	16.7	37.6	21.4
19	190	0.188	130	8	18.2	18.4	37.7	17.7
20	195	0.188	130	8.8	21.1	18.8	36.1	15.2
21	200	0.188	130	9.5	23.9	20.7	34.6	11.3
22	205	0.188	130	8.7	20.2	18.8	35.1	17.2
23	210	0.188	130	6.3	16.6	15.8	38.1	23.2
24	215	0.188	130	7.8	17	16.6	38	20.6
25	220	0.188	130	9.7	21.2	18.2	36.8	14.1

Inputs were X1 (melt temperature in°C), X2 (melt stream size in inches), X3 (type of material); outputs were weight fractions having micrometer ranges as specified by Y1 (0–53), Y2 (53–106), Y3 (106–150), Y4 (150–295), and Y5 (295–600)

Table 3. Forward prediction of the weight fractions of particles having the micrometer ranges specified by Y1 through Y5

Problem	Partial least squares	Neural networks
Forward (3 inputs, 5 outputs)	Y1: <37.19%>, max.=86.30%	Y1: <7.35%>, max.=19.75% (FNN)
		Y1: <5.17%>, max.=20.7% (GRNN)
	Y2: <68.30%>, max.=68.30%	Y2: <9.78%>, max.=32.18% (FNN)
		Y2: <10.6%>, max.=33.6% (GRNN)
	Y3: <25.85%>, max.=57.90%	Y3: <12.20%>, max.=41.97% (FNN)
		Y3: <10.8%>, max.=32.5% (GRNN)
	Y4: <20.80%>, max.=86.70%	Y4: <7.26%>, max.=24.50% (FNN)
		Y4: <6.3%>, max.=25.0% (GRNN)
	Y5: <17.6%>, max.=47.30%	Y5: <5.37%>, max.=16.00% (FNN)
		Y5: <7.5%>, max.=19.8% (GRNN)

Table 4. Reverse prediction of X1 through X3 based on weight fractions of particles having the particles having the micrometer ranges specified by Y1 through Y5

Problem	Partial least squares	Neural network
Reverse (5 inputs, 3 outputs)	X1: <37.70%>, max.=89.60%	X1: <35.0%>, max.=103.4%
	X2: <22.50%>, max.=44.90%	X2: <15.9%>, max.=30.0%
	X3: <14.80%>, max.=38.50%	X3: <3.7%>, max.=10.6%

ders in the Y1 size range can be made. The results suggest that Y1 can be increased by increasing the temperature (X1) and decreasing the melt stream size (X2) for the material (X3) equal to PE130. The neural network model predicts a value of Y1=16.5 for X1=290 °C, X2=0.05 in, X3= PE 130. Trends in the predictions show that decreasing the melt stream size at a constant temperature leads to poorer size-ranged particles. The Y1 weight fraction increased only by increasing the temperature while decreasing the melt stream size. The following specific observations were made in the computational neural network modeling: 1) decreasing X2 and increasing X1 leads to maximal Y1 for certain ranges of X1 and X2 (i.e., X1=290 °C, X2=0.05 for X3=PE 130); 2) decreasing X2 for X1<210 °C leads to a decrease in Y1; and 3) X3=PE 130 appears to be the best material for maximizing Y1 independent of X1 and X2. These observations are consistent with the experimental results obtained from current research.

Predictions from the various models are compared with experiment in Figs. 7 and 8. The GRNN and FNN predictions give correlation coefficients better than 0.98 while the PLS prediction had a poorer correlation coefficient of 0.82 (see Fig. 7). Figure 8 illustrates the predictive capability for the reverse problem. Neural networks and PLS both perform substantially worse but still provide

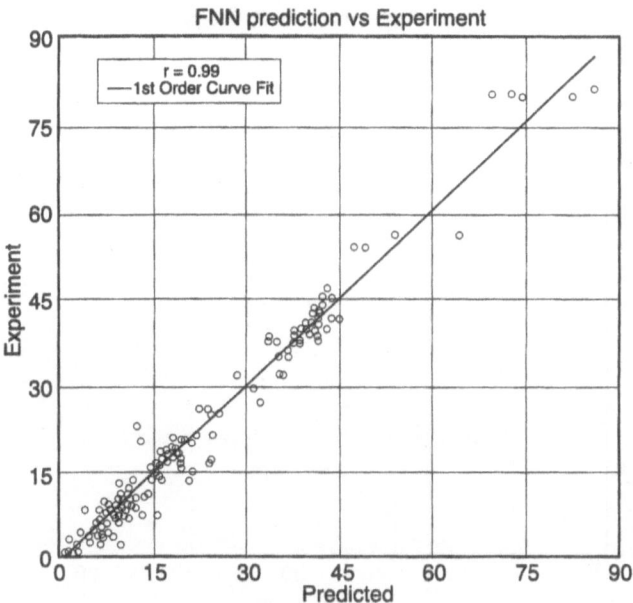

Fig. 7a. *Comparison of prediction (**forward mode**) with experiment:* Feedforward neural network (FNN) predictions described in the text

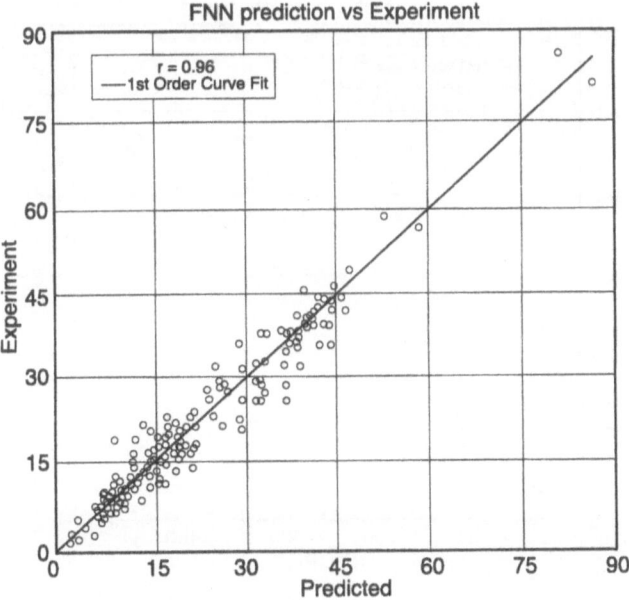

Fig. 7b. *Comparison of prediction (**forward mode**) with experiment:* GRNN predictions described in the text

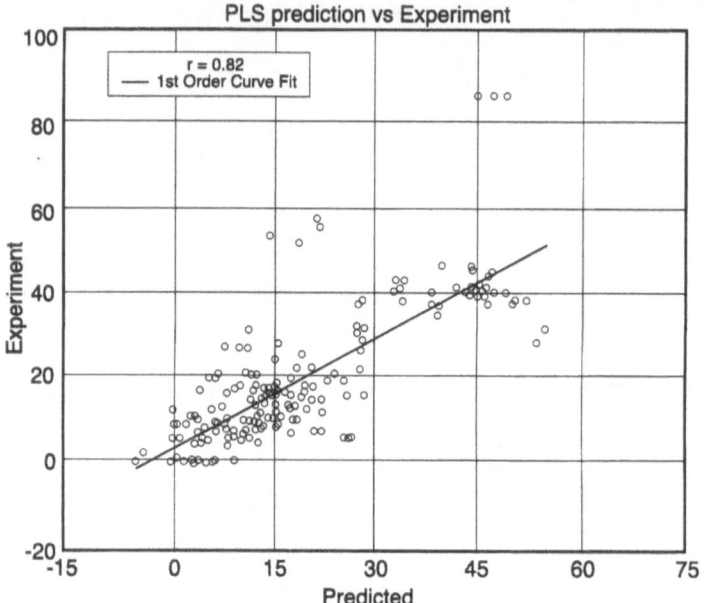

Fig. 7c. *Comparison of prediction (**forward mode**) with experiment:*
PLS predictions described in the text

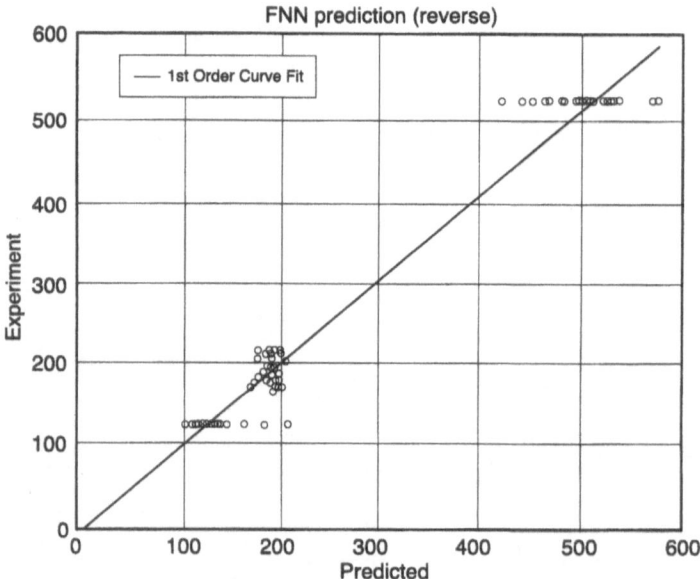

Fig. 8a. *Comparison of prediction (**reverse mode**) with experiment:*
FNN predictions described in the text

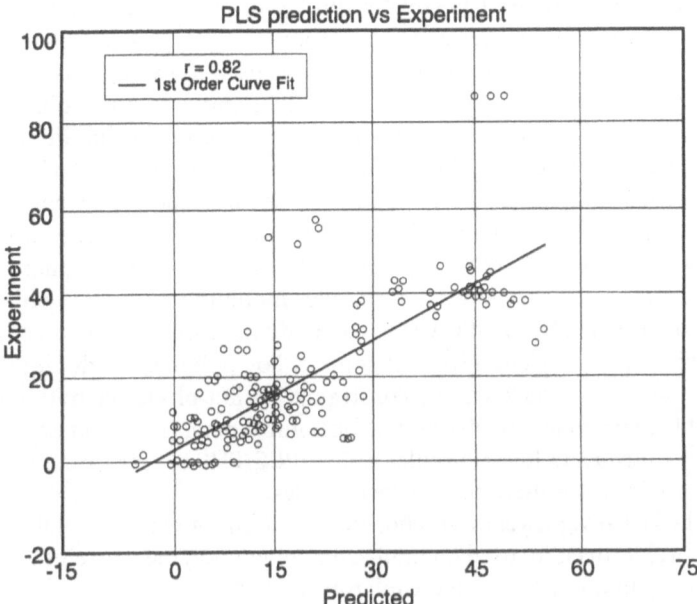

Fig. 8b. *Comparison of prediction (**reverse mode**) with experiment:*
PLS predictions described in the text

of the reverse problem, approximate values for temperature and melt stream size could be obtained for maximizing the weight fraction of particles in the 0–53 μm range. Following this reverse engineering simulation, the neural network model of the forward process was used to fine-tune the parameters. A new set of experiments has been defined by using this procedure and will be reported in a future publication. Furthermore, the results of the reverse model led to the discussions of other possible techniques for generating polymeric microparticles, leading to the development of the microdroplet approach to polymers.

The preceding discussions show that statistical and computer neural network modeling can be applied to the GAP to obtain models that can be used to predict performance of the process in the forward and reverse modes. The neural network model predicts the performance to within an average of 12% for the forward and 35% for the reverse modes, making it a useful model for expanding the GAP technology to other systems without the need to perform large and expensive experiments.

4
Microdroplet Approach to Generation and Analysis of Monodisperse Particles

In this section, we describe a new method for producing and characterizing polymer and polymer-composite particles from solution using microdroplet techniques. In addition, some of the tools developed for size characterization of liquid droplets are now being used as sensitive probes of phase separation behavior in polymer-blend microparticles generated from microdroplets of dilute polymer solution. Two-dimensional optical diffraction is used to probe phase-separation behavior of bulk-immiscible polymers in attoliter and femtoliter volumes. As we discuss further, this technique is sensitive to the presence of sub-domains with dimensions on the order of ≈30 nm, providing information on material homogeneity on a molecular scale. Under conditions of rapid solvent evaporation (i.e., very small droplets) and relatively low polymer mobility, homogeneous particles can be formed using different polymers that ordinarily undergo phase-separation in bulk preparations. Effects of polymer mobility on phase-separation behavior in microparticles are illustrated using mixtures of polyvinyl alcohols with high-mobility polyethylene glycol oligomers. For polymers with large (>10 k) molecular weights, surface energy constraints inhibit phase separation and the polymer blend particles are observed to be homogeneous to within experimental resolution. Conversely, low molecular weight PEG/PVA blends form heterogeneous sphere-within-a-sphere composite particles.

Over the last several years, an enormous amount of experimental and theoretical effort has been focused on multi-component polymer systems as a means for producing new materials on the micron and nanometer scale with specifically tailored material, electrical, and optical properties. Composite polymer particles, or polymer alloys, with specifically tailored properties may find many novel uses in such fields as electro-optic and luminescent devices [147–148], conducting materials [149], and hybrid inorganic-organic polymer alloys [150]. A significant barrier to producing many commercially and scientifically relevant homogeneous polymer blends[151–154], however, is the problem of phase separation from bulk-immiscible components in solution. This problem has been studied in detail by several different groups [155–157]. The route typically taken in trying to form homogeneous blends of immiscible polymers is to use compatibilizers to reduce interfacial tension. Recently, a number of different groups have examined phase-separation in copolymer systems to achieve ordered *meso-* and micro-phase separated structures with a rich variety of morphologies [22, 158]. For solvent-cast composites, phase separation and related morphologies depend on the time scale for solvent evaporation relative to molecular organization.

Our interest is in using small droplets (≈5–10 μm diameter) of dilute mixed-polymer solution to form homogeneous polymer composites without compatibilizers as a possible route to new materials with tunable properties. Over the last several years, advances in microdroplet production technology for work in single-molecule detection and spectroscopy in droplet streams has resulted in generation of droplets as small as 2–3 μm with a size dispersity of better than 1%. In the context of polymer particle generation, droplet techniques are attractive since particles of essentially arbitrary size (down to the single polymer molecule limit) can be produced by adjusting the size of the droplet of polymer solution, or the weight fraction of the polymer in solution. While droplet production in the size range of 20–30 μm (diameter) is more or less routine (several different on-demand droplet generators are now available commercially), generation of droplets smaller than 10 μm remains non-trivial – especially under the added constraint of high monodispersity. Small droplets (<10 μm) are especially attractive as a means for producing multi-component polymer-blend and polymer-composite particles from solution since solvent evaporation can be made to occur on a millisecond time scale, thus inhibiting phase-separation in these systems.

The primary condition for suppression of phase separation in these systems is that solvent evaporation must occur on a time scale that is faster than the self-organization times of the polymers. This implies time scales for particle drying on the order of a few milliseconds implying droplet sizes ≤10 μm (depending on solvent, droplet environment, etc.). We have shown recently that a microdroplet approach can be used to form homogeneous composites of co-dissolved bulk-immiscible polymers [29] using instrumentation developed in our laboratory for probing single fluorescent molecules in droplet streams [159, 160]. In addition to a new route to forming nanoscale polymer composites, a microparticle format offers a new tool for studying multi-component polymer blend systems confined to femtoliter and attoliter volumes, where high surface area-to-volume ratios play a significant role in phase separation dynamics.

In this chapter, we describe in some detail the basis of optical diffraction in spherical dielectric particles as a probe of material homogeneity in polymer composites, and discuss limitations of domain size (in multi-phase composites) and dielectric constant. We show how this measurement technique can be used to recover information on drying kinetics, inter-polymer dynamics, and material properties. In the following chapter, we describe results of detailed molecular dynamics modeling that can be used to connect experimental observations with microscopic dynamics within the particle as well as to suggest future experiments. The organization of this chapter on synthesis and characterization of polymer and polymer-composite particles is as follows: First, we describe our instrumentation for producing, manipulating, and characterizing polymer particles from microdroplets of solution. Next, we summarize some of the important results. Finally, we discuss some exciting possible future directions and applications.

4.1
Experimental Methods

4.1.1
Droplet Production

Several different choices exist for producing micron and sub-micron droplets of solution – each with certain trade-offs in terms of volume throughput, nominal size, and size dispersity. There are additional trade-offs associated with sampling and interrogation facility that should be considered as well. Two familiar methods of droplet production include electrospray generation and aerosol generation using a vibrating orifice coupled to a high-pressure liquid stream. In electrospray generation, a liquid stream is forced through a needle biased at approximately 1 kV (nom.). Charge-carriers induced on the surface of the liquid stream eventually come close enough during solvent evaporation for Coulomb repulsion to occur (the Taylor cone), resulting in fragmentation, or explosion, of the liquid stream. This results in a cloud of charged liquid droplets whose size can be made quite small (≤1 μm). The obvious drawback is that it is difficult to isolate individual particles for study, and size dispersity tends to be highly sensitive to experimental parameters.

Another common method of producing liquid droplets is the technique of vibrating orifice aerosol generation (VOAG). Invented by Berglund and Liu back in the 1970s, the VOAG is unmatched in terms of volume throughput (>100 nanoliters per second) and size dispersity (<0.1% depending on experimental conditions). This technique works by introducing a high-frequency (10 to 100 kHz) instability in a high-pressure liquid stream applied by a piezoelectric transducer (PZT). The resulting fragmentation of the stream produces highly monodisperse droplets that are ultimately limited by the purity of the RF signal applied to the PZT. Some disadvantages of this mode of production are that the droplets travel at high speeds (several meters/second), and are quite close together (typically no more than 3 droplet diameters apart). This makes isolation and spectroscopic interrogation of individual droplets difficult. A more serious problem is that there are significant size limitations associated with this technique. A VOAG works best at a size range of 30–50 μm, but can function down to about 12–15 μm. Because of the way droplets are produced, however, there is a concomitant increase in the RF frequency that can be problematic.

The method we have chosen in our experiments is an "on-demand" or droplet ejection device. Like a VOAG, it also uses piezoelectric transduction but at much lower frequencies. The physics of droplet production is completely different for the two methods: The VOAG operates by generating a high (and fixed) frequency instability in a liquid stream; the on-demand droplet generator functions by the application of a acoustic wave to a static solution, which forces (ejects) a droplet out of a micron sized orifice. We use Pyrex or quartz tubing that is heated, drawn, cut and polished to produce a tapered orifice that can range in size from 1 to 50 μm. The droplet sizes are comparable (usually slightly larger) to the orifice diameter and, depending on the quality of the orifice, size dispersity less than 1% can be achieved. Droplet production rates tend to be significantly lower than the aforementioned techniques. Ultimately, droplet rates are limited by piezoelectric relaxation times (≈10 kHz); practically, however, under conditions of high monodispersity, droplet rates are typically much lower (20–100 Hz). The advantages of this technique are small size and on-demand production that makes single droplet/particle manipulation and interrogation straightforward.

4.1.2
Diffraction-Based Probes of Material Homogeneity

Our primary experimental tool for probing phase-separation behavior and material homogeneity in polymer composites is essentially an interferometric technique that has been used for a number of years as a method for sizing liquid microdroplets (size range ≈ 2–20 µm. Recently, this measurement technique has been used to recover information on drying kinetics, inter-polymer dynamics, and material properties such as dielectric constant. The basis of the technique involves illuminating a dielectric sphere with a plane-polarized laser to produce an inhomogeneous electric field intensity distribution, or grating, within the particle that results from interference between refracted and totally-internally-reflected waves within the particle. The angular spacing between intensity maxima, as well as the intensity envelope is a highly sensitive function of particle size and refractive index (both real and imaginary parts). Unlike conventional microscopy approaches with diffraction-limited ($\approx \lambda/2$) spatial resolution, two-dimensional diffraction (or, angle-resolved scattering) is sensitive to material homogeneity on a length scale of $\approx \lambda/20$ or about 20–30 nm for optical wavelengths. This dimension is comparable to single-molecule radii of gyration for relative large molecular weight (>100 k) polymers, and thus provides molecular scale "resolution" of material homogeneity in ultra-small volumes (≈ 1–100 femtoliters).

Light scattering from micron-sized spherical dielectric droplets or particles [161] has been used for a number of years as a method of sizing [162–164] and analyzing various physical and chemical properties [165, 166]. Various light scattering techniques from spherical droplets have been very well characterized for particle sizing and refractive index determination [167–170]. Recently, use of 1- and 2-dimensional angle-resolved elastic scattering has been utilized as a tool for characterizing in situ polymerization in microdroplets [171, 172], and probing multi-phase [173, 174] and homogeneous [29] composite particles. The basis of the technique involves illuminating a dielectric sphere with a plane-polarized laser to produce an inhomogeneous electric field intensity distribution, or grating, within the particle that results from interference between refracted and totally-internally-reflected waves within the particle. The angular spacing between intensity maxima, as well as the intensity envelope is a highly sensitive function of particle size, and refractive index (both real and imaginary parts). As shown in Ref. 29, the fringe contrast (and intensity fluctuations along an individual fringe) is also very sensitive to material homogeneity on a length scale of $\approx \lambda/20$ or about 20–30 nm for optical wavelengths. In our experimental configuration, this intensity grating is projected in the far-field using (f/1.5) collimating optics and detected with a CCD camera [175]. The scattering angle (center angle and width) is established by means of an external calibration, and is used for high-precision Mie analysis of one-dimensional diffraction data.

For optical diffraction studies, individual particles were studied using droplet levitation techniques. Details of the apparatus and CCD calibration procedure are described in Ref. 175. The nominal scattering angle was 90 degrees with respect to the direction of propagation of the vertically polarized HeNe laser, and the useable full plane angle (defined by the f/1.5 achromatic objective) was 35 degrees. The CCD (SpectraSource Instruments) was thermoelectrically cooled and digitized at 16 bits. Details of the droplet generator used are described in Ref. 160. Aqueous solutions were handled by simply loading the Pyrex tip by vacuum aspiration and re-installing into the generator. For the work done on co-dissolved polymers in tetrahydrofuran (THF), the entire droplet generator chamber and ballast reservoir were backfilled with THF, and the tip was loaded with the polymer solution of interest.

In the work reported here, we used a 60 Hz AC potential for particle levitation. Due to charge-to-mass instabilities, particles with diameters smaller than about 2-μm were not trapped effectively at 60 Hz. This limitation can be overcome by using higher frequencies at the expense of additional experimental complexity [176]. A more serious issue is the information content in the data for particles smaller than about 2 μm. Geometric limitations on the range of measured scattering angles also place a lower size limit on the particles that can be interrogated with this technique. Since the fringe spacing increases with decreasing particle size, we estimate that diffraction from particles smaller than about 1.5 μm would be difficult to analyze quantitatively with this approach since the angular range (35 degrees) is fixed. In addition, the integrated scattered light intensity decreases approximately as the square of the diameter in this size range – compensation for this effect by increasing laser power often adds the complication of photon pressure [176], thus exacerbating the problem.

Figure 9 shows some examples of one-dimensional diffraction data for polyethylene glycol particles (10 k avg. molecular weight) produced from aqueous solution and Mie theory matches to the data. The size can be tuned nearly arbitrarily by adjusting the droplet size and weight fraction of the polymer. The points are measured scattering intensities as a function of laboratory angle, and the smooth curves are Mie theory matches to the data optimized with respect to particle size and both real and imaginary parts of the refractive index. Note that the Mie analysis is not a "fit" to the data in the sense of a non-linear least squares optimization of parameters or coefficients associated with some predefined function. The three independent factors that define the scattering pattern – size, Re(n), and Im(n) – are systematically varied to find the best possible match to the experimental data by locating the minimum in a 4-dimensional error function.

4.2
Experimental Observations and Discussion

Figure 10 shows a family of 2-D slices (Re(n) varied and Im(n) fixed) of a typical error surface obtained from data acquired for a PEG particle. From exhaustive analysis of these error surfaces from many different sized particles, we find absolute size uncertainties to be between 2 and 5 nanometers, and the uncertainty in Re(n) to be between 10^{-3} and 5×10^{-4}. For materials with a low molar absorptivity (typical of most dielectric liquids), Im(n) is correspondingly small – on the order of 10^{-5} to 10^{-7}. At these values, there is very little (if any) effect on the match to data by varying Im(n). For many polymers (e.g. polyvinyl chlorides) however, this is not the case and Im(n) can be as large as 10^{-3}. At this order of magnitude, Im(n) does indeed influence the Mie analysis of the data.

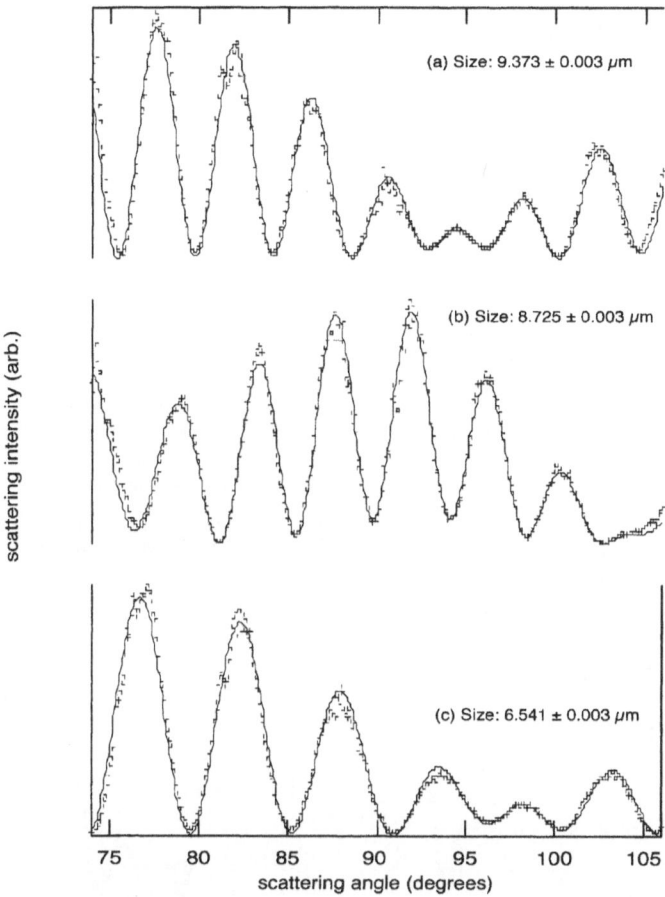

Fig. 9. Mie theoretical matches to one-dimensional (row) slices of measured two-dimensional diffraction patterns from polyethylene glycol particles produced from microdroplets of solution with different PEG weight fractions

A key issue in forming homogeneous composites from co-dissolved bulk-immiscible polymers from solution is that the droplet evaporation rate must be faster than the polymer self-organization time scale. Since the time scale for solvent evaporation is proportional to $1/(r^{3/2})$, (r is the droplet radius), the most straightforward way to satisfy this condition is to make droplets smaller. [159] Alternatively, one can modify the atmosphere around the droplet (temperature, different bath gases, etc.) to accelerate particle drying [160]. As part of our effort in developing single-molecule isolation and manipulation methodologies in droplet streams, we have been able to produce (water) droplets as small as 2 – 3 microns in diameter with ≈1% size dispersity. Figure 11 shows an experimental characterization of droplet evaporation on a millisecond time scale using a high-speed frame transfer CCD camera. Angular scattering data was acquired at suc-

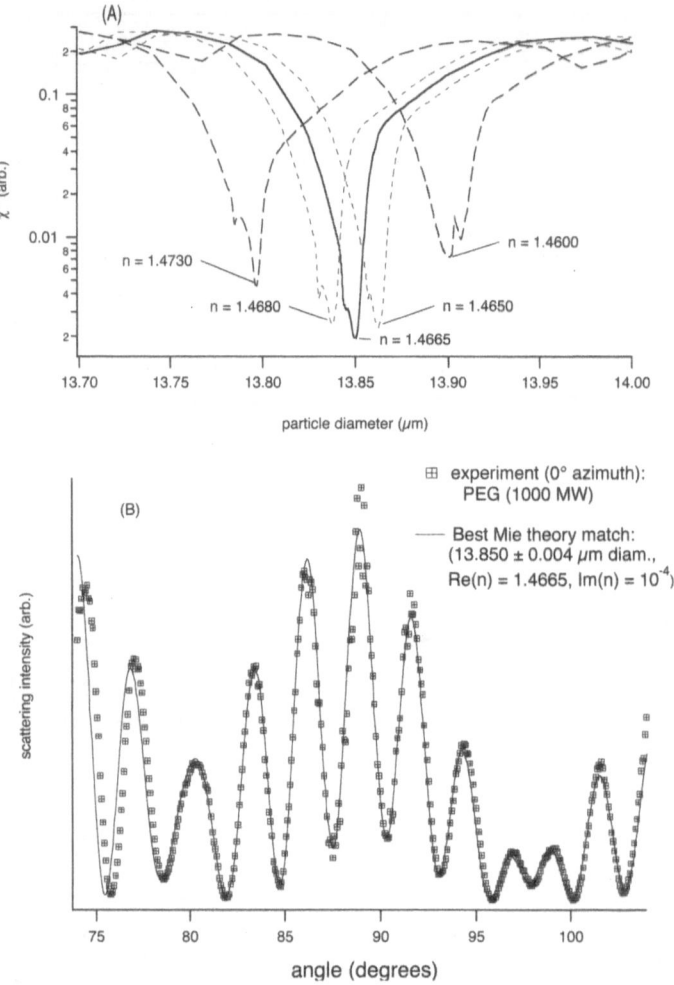

Fig. 10. Two-dimensional slices of 4-dimensional error surface (varying Re(n)) for Mie theory match to diffraction from PEG particle (Im(n) fixed). The lower trace shows the best match to the experimental scattering data

cessive 1-ms intervals during the 5-ms transit of the droplet through the laser beam. The deviation of the third and fifth points from the model most likely originate from a "phase-aliasing" due to particle motion induced from photon pressure. For a 3-μm (nom.) droplet, approximately 95% of the initial droplet volume is lost due to evaporation in about 5 ms. For a polymer molecule with a modest molecular weight of say, 10^4, the diffusion coefficient is on the order of 10^{-9} cm^2/s. In this case, diffusional motion (of the polymer center-of-mass) is negligible on the time scale of solvent evaporation.

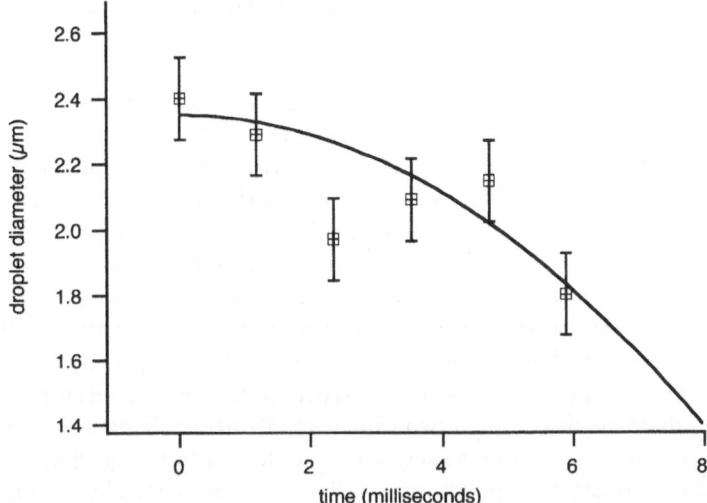

Fig. 11. Characterization of rapid evaporation from a 2.6 μm diameter water droplet using angular scattering and a high-speed frame transfer CCD camera. Since the CCD pixel transformation parameters were not precisely defined in these experiments the errors are estimated to be on the order of the diffraction limit

We have recently shown that the presence of phase-separated structures in polymer-blend microparticles can be indicated qualitatively by a distortion in the two-dimensional diffraction pattern. The origin of fringe distortion from a multi-phase composite particle can be understood as a result of refraction at the boundary between domains of different polymers, which typically exhibit large differences in refractive index. Thus, the presence of separate sub-domains introduces optical phase shifts and refraction resulting in a "randomization" (distortion) in the internal electric field intensity distribution that is manifested as a distortion in the far-field diffraction pattern.

4.2.1
Domain size 'resolution' in 2-D diffraction

A critical question in the analysis of two-dimensional diffraction patterns from composite microparticles is the minimum domain size that can be detected with this technique. Light scattering and resolution analysis of Fabry-Perot interferometers suggest that refractive index discontinuities (phase-separated domains) with a length scale of about $\lambda/20$ ($\approx$25–30 nm), where λ is the laser wavelength, are required to produce a measurable distortion in the diffraction pattern. If true, this "resolution" is comparable to radii of gyration for most large molecular weight polymers, thus providing a method of looking "within" a composite particle and probing material homogeneity on a molecular scale. Here we examine in some detail the question of domain size and number density of phase-separated sub-domains within a host particle. Empirically and theoretically, we find that this rule-of-thumb provides a reasonably good approximation.

In this subsection, we address empirically the issue of domain size resolution in two-dimensional optical diffraction of polymer-blend and polymer-composite microparticles. The question can be phrased in two parts: (1) What is the smallest phase-separated sub-volume (domain) of the particle that will manifest its presence as a distortion in the two-dimensional diffraction pattern? (2) What is the sensitivity to number density or relative weight fraction? That is, if two polymers phase-separate in a microparticle, at what relative weight fraction will one to be able observe the phase separation (0.1, 1, 10% etc.) through distorted diffraction patterns. These questions are not trivial to address by first-principles electrodynamics, although a perturbative volume-current method has recently been employed to simulate distortion of two-dimensional diffraction patterns from binary particles. These calculations suggest that, under the right experimental conditions, very low relative weight fractions (<1%) are required to observe distortion, with domain size resolution on the order of 20–40 nm depending on the relative difference in refractive index between host and guest material.

Experimentally, we find good agreement with domain size resolution but somewhat poorer agreement in number density (relative weight fraction) requirements for diffraction distortion. We examined diffraction from polyethylene glycol host particles doped with varying weight fractions of ceramic nanoparticles (Al_2O_3, TiO_2) and 14-nm latex beads. The Al_2O_3 and TiO_2 have nominal sizes of 45 and 28 nm respectively but their refractive indices (1.57, and 2.1 respectively) are such that the two particles introduce approximately the same optical phase shift. Indeed, the dopant ceramic particles produce measurable distortion in the diffraction patterns (quantitatively described by Fourier transform of individual diffraction fringes), but only at relative weight fractions >5%. Interestingly (see Sect. 4.2.2), adding 14-nm latex beads to PEG host particles did not result in distortion of the diffraction pattern at relative weight fractions up to 50%, but do significantly modify the refractive index.

4.2.2
Material Properties and Particle Dynamics

Another important aspect of our approach is the ability to obtain very precise information on the refractive index, which provides a way of characterizing material properties and particle dynamics that is not possible with conventional microscopy techniques. For example, one can directly measure kinetics of particle drying by monitoring the refractive index change as the solvent evaporates. In general, we find that the particle drying kinetics have two distinct solvent evaporation regimes. When the droplet is first ejected, the size decreases rapidly with most of the solvent evaporation taking place within the first 10 to 100 milliseconds. This is followed by a much slower ("wet particle") evaporation regime where the particle continues to lose solvent on a time scale of several minutes (dependent on size). Interestingly, in the slow-evaporation regime, we generally observe that the volume change expected from solvent loss (assuming an ideal solution) is *not* accompanied by the expected corresponding change in particle size. Put another way, we observe a much smaller change in size than would be expected from the change in refractive index (solvent loss) if the particle were able to respond to solvent removal in the same way as an ideal solution. This suggests that, in most cases, the particle forms a semi-rigid matrix allowing trapped residual solvent to escape by diffusion.

We examined two bulk-immiscible polymers (polystyrene and polyvinyl chloride) that are soluble in THF [29]. We have not characterized evaporation rates of pure THF droplets, but estimate on the basis of vapor pressure differences relative to water, that evaporation is roughly a factor of five times more rapid (for a given droplet size). To provide some kind of relative time scale, a water droplet with an initial diameter of 10 µm will evaporate to ≈1 µm diameter in about 5 ms in a dry Argon atmosphere [160].

Figure 12 shows examples of PVC/PS particles formed from an (a) 8-µm diameter droplet, and (b) a 35-µm diameter droplet of dilute (≈1% total polymer weight fraction) PVC/PS/THF solution. The size threshold (for this system) for producing homogeneous particles is about 10 µm. For larger droplets with correspondingly longer drying times, we observed that the particles were inhomogeneous. There is compelling evidence for material homogeneity at a molecular length scale for the particle represented in Fig. 12-a: fringe uniformity, quantitative agreement with Mie calculations, and a refractive index (related to material dielectric constant) that is intermediate between the two pure materials [147]. It should be noted that for heterogeneous particles, while the diffraction fringes are highly distorted, there remains some definite two-dimensional structure. This implies some uniformity and order of subdomains, however, inversion of this type of data to extract such information is not trivial. This problem is currently under investigation. (See also Ref. 173).

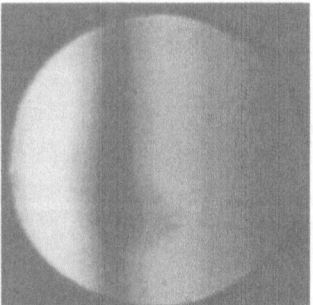 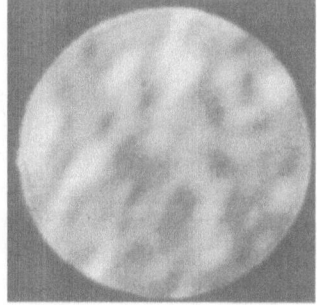

Fig. 12. Two-dimensional diffraction data from 50:50 w/w PVC/PS blend particles produced from an 8 µm diameter droplet (*left*), and a 35 µm diameter droplet (*right*)

Figure 13 shows time-resolved results of particle size and refractive index for a 2.4-μm PVC/PS (50:50 w/w) blend particle. Note that the particle continues to "dry" on a time-scale of several minutes, but remains homogeneous throughout the measurement sequence. The nominal index of pure THF is 1.41 and the measured (steady state) final index of the particle is 1.527. These limiting values suggest that, for the first data point in Fig. 13, the particle is 22% by volume THF. Assuming that the last data point represents a fully dry particle (in good agreement with estimates based on refractive indices for pure PS and PVC), one would have expected a 60% *decrease* in particle size accompanying the loss of residual THF. This is clearly not observed, indicating that the particle has formed a fairly rigid matrix that compresses only slightly (≈3%) during the exposure sequence.

The data shown in Fig. 13 also illustrates the "tunable" nature of a material property in PVC/PS composite particles – namely the dielectric constant manifested in the refractive index. Both Re(n) and Im(n) for the polymer-blend microparticles are intermediate between the values determined for pure single-component particles (PVC: Re(n) =1.4780, Im(n) =10^{-3}; PS: Re(n) =1.5908, Im(n) =2×10^{-5}) and can be controlled by adjusting the weight fractions of polymers. Interestingly, the measured refractive index for composite particles is very close to estimates obtained from a simple mass-weighted average of the two species.

In order to develop some insight into the structure, and properties of polymer blend particles, we have also investigated this problem using molecular dynamics simulation tools. Using classical molecular dynamics techniques (discussed in detail in the following chapter), we have examined polymer nanoparticles of

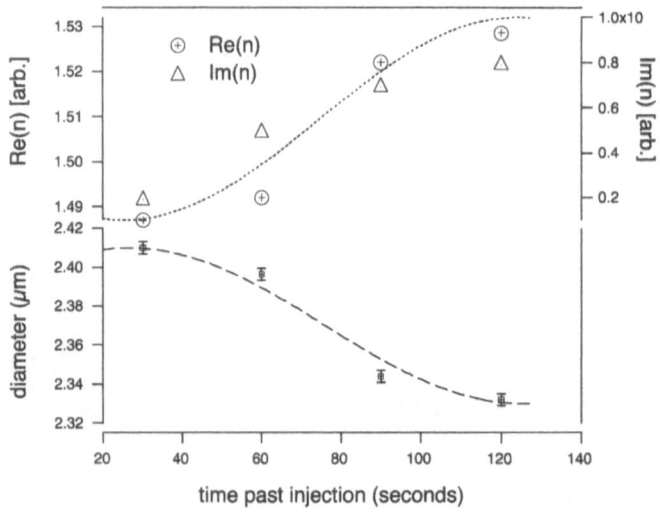

Fig. 13. Particle size and refractive index for a PVC/PS composite particle (50:50 w/w) as a function of time past injection

varying size (up to 300,000 atoms), chain-lengths (between 50–200 monomers), and intermolecular interaction energy allowing the systematic study of size-dependent physical properties and time dependence of segregation/equilibration of these particles.

Figure 14 shows molecular dynamics simulations of a stable polymer-blend particle (10-nm diameter) composed of immiscible components [177]. The leftmost particle remains homogenous throughout a broad temperature range. For phase separation to occur (right), an enormous amount of thermal energy must be supplied to overcome the surface energy barrier. This result agrees qualitatively with the observation that homogenous blends of bulk-immiscible polymers can be formed in spherical microparticles. The composite-particle was calculated to have a *single* melting temperature of 190 K and glass transition temperature of 90 K, which is different than either of the polymer components (T_m =218 K, T_g =111 K for light and T_m =162 K, T_g =81 K for dark). The segregated particle has two melting points and glass transition temperatures that correspond to within 10 K of the individual components.

Formation of homogeneous polymer-blend composites from bulk-immiscible co-dissolved components using droplet techniques has two requirements. First, solvent evaporation must occur on a relative time scale compared to polymer translational diffusion. Secondly, mobility must be low enough so that, the polymers cannot overcome the surface energy barrier and phase-separate, once the solvent has evaporated. We have shown definitively the effects of droplet size and solvent evaporation, and the second requirement is almost always satisfied even for modest molecular weight polymers. In order to explore effects of polymer mobility in more detail, we looked at composite particles of PEG oligomers (MW 200, 400, 1000, and 3400) with medium molecular weight (14 k) atactic polyvinyl alcohol (PVA). This system allowed us to systematically examine the

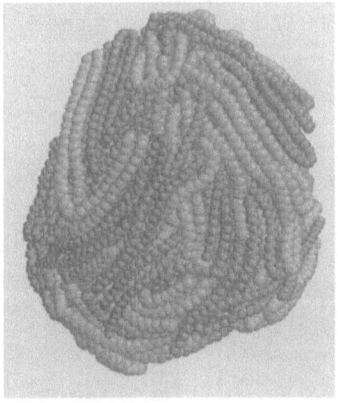

Fig. 14. Calculated structures for homogeneous (*left*) and phase-separated (*right*) 10-nm polymer blend particles

phase separation behavior where one component (PEG) had substantially different viscosities (specified as 4.3, 7.3, and 90 cStokes at room temperature for PEG
[200], PEG [400], and PEG [3400] respectively).

We observe that the higher molecular weight PEG polymer-blend particles
are homogeneous as determined from bright-field microscopy, optical diffraction, and fluorescence imaging. Blend-particles prepared with the 200-molecu-
lar weight PEG were observed to form sphere-within-a-sphere particles with a
PVA central core. Figure 15 shows diffraction data acquired particles at different
times from a 10-μm diameter PEG[200]/PVA[14 k] (80:20 w/w) particle. As
shown in the early frame, the particle is initially homogeneous. On a time scale
of about 10 minutes, the composite particle undergoes phase separation into an
inhomogeneous particle as evidenced by the fringe distortion. Interestingly, the
structure in the two-dimensional diffraction data for this system is much different than those observed for large phase-separated PVC/PS particles that pre-
sumably coalesce into sub-micron spheroidal domains [178]. Based on the imaging data shown in Fig. 7 of Ref. 179, the PEG[200]/PVA[14 k] particle forms
spherically symmetric (sphere-within-a-sphere) heterogeneous structures,
which should also produce well-defined diffraction fringes [180–182].

Our interpretation of this data is that diffusional motion of the PVA core in the
PEG host particle, combined with rotational diffusion of the particle breaks the
spherical symmetry, thereby introducing distortion in the diffraction pattern.
This observation is entirely consistent with our model of polymer-composite formation where heterogeneous particles may be formed provided that the mobility
of one of the polymers is low enough to overcome the surface energy barrier.
Composite particles formed from the higher molecular weight PEG (>1000) form
homogeneous blend particles with PVA. The time scale for transitions from a homogeneous to inhomogeneous particle provides an estimate (at least to an order
of magnitude) of the diffusion coefficient of the light component *within* the particle. This is a number that can be connected directly to molecular dynamics simulations. By equating the average diffusion distance $r = (6Dt)^{1/2}$ to the particle radius
(6 μm) and using a value of $t = 10$ min (600 s), we estimate a value of $D = 10^{-10}$

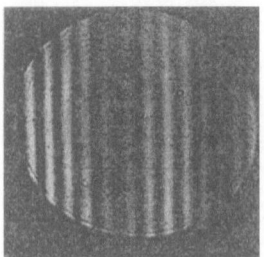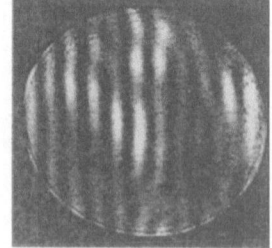

Fig. 15. Two-dimensional diffraction data acquired particles at different times from a 10-μm
diameter (nom.) PEG[200]/PVA[14 k] (80:20 w/w) particle

cm^2/s, which is consistent with recent molecular modeling results (see the following chapter). Composite particles formed from the higher molecular weight PEG (>1000) form homogeneous composite particles with PVA.

5
Molecular Dynamics Studies of Polymer Nano-Particles

Molecular dynamics (MD) is an invaluable tool to study structural and dynamical details of polymer processes at the atomic or molecular level and to link these observations to experimentally accessible macroscopic properties of polymeric materials. For example, in their pioneering studies of MD simulations of polymers, Rigby and Roe in 1987 introduced detailed atomistic modeling of polymers and developed a fundamental understanding of the relationship between macroscopic mechanical properties and molecular dynamic events [183–186]. Over the past 15 years, molecular dynamics have been applied to a number of different polymers to study behavior and mechanical properties [187–193], polymer crystallization [194–196], diffusion of a small-molecule penetrant in an amorphous polymer [197–199], viscoelastic properties [200], blend [201,202] and polymer surface analysis[203–210]. In this article, we discuss MD studies on polyethylene (PE) with up to 120,000 atoms, polyethylproplyene (PEP), atactic polypropylene (aPP) and polyisobutylene (PIB) with up to 12,000 backbone atoms. The purpose of our work has been to interpret the structure and properties of a fine polymer particle stage distinguished from the bulk solid phase by the size and surface to volume ratio.

Recently, an experimental technique was developed for creating polymer particles of arbitrary composition and size [3, 28, 29] as discussed in Sect. 4. In the experiment, we used previously developed instrumentation generating and characterizing droplet streams with small (1–2 um) average diameter and monodispersity for probing single molecules in solution [159, 211]. This technique makes the initial volume of dilute solution sufficiently small so that the solvent evaporates on a short time scale of a few milliseconds. For micro- and nano-scale generated polymer particles, the refractive index is consistent with bulk (nominal) values, and the level of agreement with Mie theory indicates that the particles are nearly perfect spheres [175]. Figure 16(a) shows a schematic image of the generation of these polymer particles; an optical image of typical particles formed from the experimental droplet generator is shown in Fig. 16(b). Those particles in the nano- and micrometer size range provide many of the unique properties due to size reduction to the point that critical length scales of physical phenomena become comparable to or larger than the size of the structure. Applications of such particles take advantage of high surface area and confinement effects, which leads to nanostructures with properties that differ from conventional materials. An example of the dependence of physical properties on size is the melting point for gold which, as is well known, decreases dramatically as the size decreases [212]. One of the major reasons for this phenomenon is the high surface area and hence reductions of the non-bonded interactions for the sur-

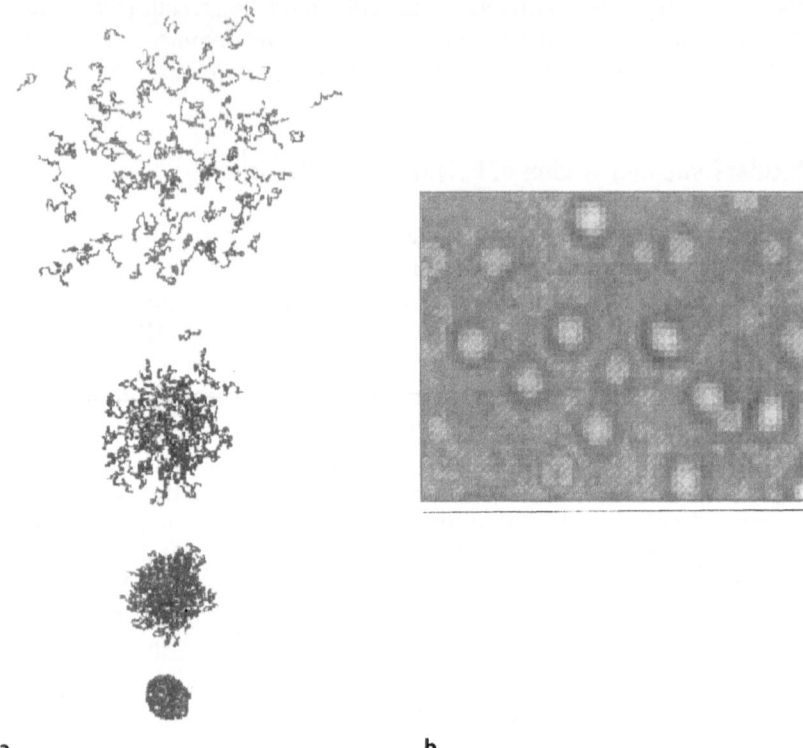

a   b

Fig. 16. a A molecular dynamics simulation was used to create a schematic image of polymer nano- and macro-scale particles generated from the submicron liquid droplets experiment. The solvent molecules (typically H_2O and THF in the experiment, not considered in the simulation) are streaming away as the polymer molecules are left behind collapsing into a nano or micrometer particle. Within the experimental time scale, the solvent evaporation was completed before the particles collected on a substrate. **b** Bright field image (inverted grayscale) of 2 μm diameter of PE particles

face layer. Clearly, such changes offer extraordinary potential for developing new materials in the form of bulk composites and blends that can be used for coatings, opto-electronic components, magnetic media, ceramics and special metals, micro- or nano-manufacturing, and bioengineering [213].

A computational algorithm for generating and modeling polymer particles for our simulations was developed to construct particles that are as similar as possible to the experimentally created polymer particles. We have examined a variety of PE nano-scale particles, allowing the systematic study of size-dependent physical properties of these particles [214]. The models have been well tested and shown to provide realistic representation of the structure and vibrational spectroscopy of a number of polymer systems: harmonic/Morse potentials for the bond stretches, harmonic potential for bending between two bonds, a truncated

Fourier series for the torsional potential, and Lennard-Jones 6–12 potentials for the non-bonded interactions (both chain-chain and intra-chain) [193, 215–218].

5.1
Model Development

The molecular dynamics method is well known and has been reviewed in several papers [219, 220]. Basically, one needs to solve Hamilton's equations or any other formulation of the classical equations of motion, starting from some initial positions and momenta of all of the atoms in the system and propagating the solution in a series of time steps. In our MD simulations, the integration of the equations of motion were carried out in Cartesian coordinates, thus giving an exact definition of the kinetic energy and coupling. The classical equations of motion were formulated using our geometric statement function approach [221], which significantly reduced the number of mathematical operations required.. These coupled equations were solved using novel symplectic integrators developed in our laboratory which conserve the volume of phase space and robustly allow integration for virtually all time scales [222].

5.1.1
Molecular Hamiltonian and Potential Energy Functions

For simplicity of the PE model, we have collapsed the CH_2 and CH_3 groups into a single particle (bead) of mass m=14.5 amu. By neglecting the internal structure of those groups, we reduce the number of equations of motion for the system, saving a significant amount of computational time. By integrating Hamilton's equations of motion using the molecular Hamiltonian written in Cartesian coordinates the MD simulations compute the momenta and coordinate of each atom in the system as a function of time The total molecular Hamiltonian is composed of the kinetic and potential energy of the system,

$$H = \sum_{i=1}^{N} \frac{p_i^2}{2m} + \sum V_{2b}(r) + \sum V_{3b}(\theta) + \sum V_{4b}(\tau) + \sum_{i=1}^{N-3} \sum_{j \geq i+3}^{N} V_{Nb}(r_{ij}),$$
(11)

where N is a total number of atoms and p_i is the Cartesian momentum of the ith atom. The r, q and t are the internal coordinates for the interatomic distance, the bending angle between three consecutive atoms, and the torsional angle between four consecutive atoms, respectively. For the two-body bonded lengths and the three-body bending angles, simple harmonic potential functions are used,

$$V_{2b}(r_{i,i+1}) = \frac{1}{2} k_r (r_{i,i+1} - r_0)^2$$
(12)

$$V_{3b}(\theta_{i,i+1,i+2}) = \frac{1}{2} k_\theta (\theta_{i,i+1,i+2} - \theta_0)^2$$
(13)

where k_r and k are the force constants parameters. The r_0 and θ_0 are the equilibrium value of the bond length and the angle formed by the three atoms of interest in a particular bond. The torsional potential for the four body angles is given by

$$V_{4b}\left(\tau_{i,i+1,i+2,i+3}\right) = 8.77 + \alpha \cos \tau_{i,i+1,i+2,i+3} + \beta \cos^3 \tau_{i,i+1,i+2,i+3} \tag{14}$$

The four-body potential term has two parameters that can be fitted to give a desired barrier height and rotational isomer energy difference. The torsional term for PE was developed by Boyd and gives a cis barrier of 16.7 kJ/mol and a *gauche-trans* energy difference of 2.5 kJ/mol [45]. The last term of the potential energy function in Eq. (11) is the non-bonded, two-body interaction separated by a sequence of three atoms or more (In our MD studies, the distance for cutting off the interaction is set at 10 C.) The function for the interaction between any two atoms not directly bonded together (1–4 types and larger, $j\ i+3$) is represented by,

$$V_{Nb}\left(r_{i,j}\right) = 4\varepsilon \left[\left(\frac{\sigma}{r_{i,j}}\right)^{12} - \left(\frac{\sigma}{r_{i,j}}\right)^{6} \right] \tag{15}$$

where ε and σ represent the Lennard-Jones parameters. The above potential energy function [Eqs. (2–5)] yields good spectroscopic, thermodynamic, and kinetic data, and also provides the atomistic details of temperature-dependent phase transitions. The parameters for the potential energy functions of Eq.(12)-(55) are shown in Table 5.

Table 5. Potential parameters

Stretch	$V_{2b}=\frac{1}{2}k_r\,(r_{i,i+1} - r_0)^2$	
C-C	r_0	k_r
	1.53 C	2651 kJ/mol
Bending[a]	$V_{3b}=\frac{1}{2}k_\theta\,(\cos\theta_{i,i+1,i+2} - \cos\theta_0)^2$	
C-C-C	θ_0	k_θ
	113°	130.1 kJ/mol
Torsion	$V_{4b}=8.77+\alpha\cos\tau_{i,i+1,i+2,i+3}+\beta\cos^3\tau_{i,i+1,i+2,i+3}$	
C-C-C-C	α	β
	–18.41 kJ/mol	26.78 kJ/mol
Nonbond	$V_{NB}=4\varepsilon[(\sigma/r_{ij})^{12} - (\sigma/r_{ij})^{6}]$	
	ε	σ
$CH_2...CH_2$	0.494 kJ/mol	3.98 Å
$CH_3...CH_3$	0.837 kJ/mol	4.14 Å

[a] For -CH- beads in PIB, the equilibrium angle, θ_0, is set at 109.47 °

5.1.2
Modeling of PE Polymer Particles

We have developed an efficient method to obtain the desired size of particles to model production of polymer particles for our MD simulations [216]. The procedure starts by preparing a set of randomly coiled chains with a chain length of 100 beads by propagating a classical trajectory with a perfectly planar all-*trans* zig-zag initial conformation and randomly chosen momentum with a temperature of 300 K [224–227]. The trajectory is terminated at 200 ps, and position and momenta of the chain are saved. By repeating this process, a desired set of randomly coiled chains is prepared. From the set, six chains are selected and placed along the Cartesian axis. To create a collision at the Cartesian origin, these chains were propelled with an appropriate amount of momentum. Then, a particle consisting of the six chains was annealed to a desired temperature and rotated through a randomly chosen set of angles in three-dimensional space, creating a homogeneous particle in agreement with the experiment. Another set of six chains was propelled at the particle to create another particle with 600 more atoms. This process is continued until the desired size is achieved for our study. Figure 17 illustrates the process to generate initial conditions for polymer particles.

We also generated the PE particles with various chain lengths (50 and 200 beads) based on the initial configurations of the generated amorphous PE with a chain length of 100 beads. To do this, we simply expanded the space of the particle with a chain length of 100 beads by multiplying the initial configuration by two in Cartesian coordinates. Then, a new atom was inserted between the two atoms of the particles. Thus, the total number of atoms was doubled and the created particle consisted of a chain length of 200 beads. For the particle with a chain length of 50 beads, the two, three and four-body bonded interaction were simply turned off every 50 beads. Finally, classical trajectories were propagated with randomly chosen momenta until the density reached 0.7 g/cm^3 and an-

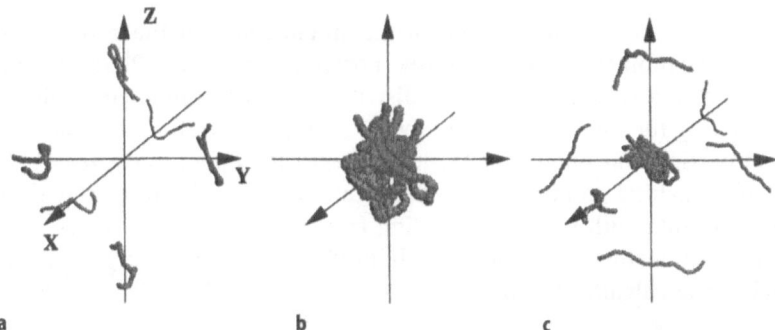

Fig. 17. Creation of polyethylene droplets for initial conditions: **a** randomly coiled 6 chains with chain length of 100 beads, **b** a particle for 600 atoms with those 6 chains, and **c** the particle and a new set of 6 chains. Before the set was slung, the particle was randomly rotated.

nealed to a desired temperature. At a start of those trajectories, the density for the particle is low (0.05 g/cm^3) and the temperature is very high (1000 K), since the space of the original particle was expanded. Using this scheme, we efficiently created large number of atoms with various chain lengths without propelling sets of six chains.

5.1.3
Mutation of Polymer Particles

A desired size of polyethylproplyene (PEP), atactic polypropylene (aPP) and polyisobutylene (PIB) particles can be efficiently modeled by "mutating" the configuration of PE particles. Expanding the space of the PE particle by multiplying the initial configuration by 1.2 in Cartesian coordinates, we insert a new atom at a desired position. For example, in the mutation of PEP from PE, 25 new atoms are placed for every PE chain at the position where the distance between a backbone (4i-2) and the inserted atom is the -CH2-CH3 equilibrium distance. The potential bonded and non-bonded interactions between the backbone and the inserted atoms are adiabatically turned on using a switching function, $SW(t)$. The switching function used for this study is,

$$SW(t) = \tanh(9.7786 \times 110^{-3} \, t) \tag{16}$$

where t is time (ps). The Hamiltonian for the mutated system can be written,

$$H' = H + SW(t) \times V_{\text{mutate}} \tag{17}$$

where V_{mutate} is the potential energy functions for the inserted atoms. The potential functions and parameters used for the polymer particles studied here are given in Table 5.

5.1.4
Trajectories

At the start of the MD simulations for the mutated system, the initial values of the Cartesian momenta were randomly chosen in phase space. To determine the initial momenta, we chose uniformly distributed random numbers on the interval [−1/2,1/2], then multiplied these by an appropriate scaling constant. After a desired particle was obtained, classical trajectories were propagated for 50 ps above bulk melting point, and then the system was annealed by scaling the Cartesian momenta with a constant scaling factor until the temperature reached 10 K to find a steady state of the amorphous PE particles. The temperature of the particle T was calculated from

$$\frac{3}{2} N k_B \langle T \rangle = \left\langle \sum_{i=1}^{n} \frac{P_i^2}{2m} \right\rangle \tag{18}$$

where k_B is the Boltzmann constant and N is the total number of atoms. The details of the annealing procedure are explained in Sect. 5.1.6.

To obtain the average values of properties of the particles at a fixed temperature and examine dependence of the conformations on temperature, we used Nose-Hoover chain (NHC) constant temperature molecular dynamics [228]. The initial configurations of the steady state of the amorphous PE particle were used at a start of the NHC simulations; the initial values of the Cartesian momenta were given random orientation in phase space with magnitudes chosen so that the total kinetic energy was the equipartition theorem expectation value [229]. The NHC quasi-Hamiltonian for this system can be written,

$$H'\left(p,q,p_\eta,\eta\right)=T\left(p_1\cdots p_N\right)+V(q_1\cdots q_N)+\sum_{i=1}^{\Omega}\frac{p_{\eta_i}^2}{2M_{\eta_i}}+N_f k_B T\eta_1+\sum_{i=2}^{\Omega}k_B T\eta_1 , \quad (19)$$

where η, p_η, and M_η are the position, conjugate momentum, and mass of the ith thermostat and W is the number of thermal baths. In this work, we employed W=3 (sufficient to ensure ergodicity). Also, k_B is the Boltzmann constant, N_f is the number of molecular degrees of freedom in the system, and T is a desired temperature. The position η and momentum p_η are set to zero at $t=0$. After we propagated NHC trajectories 10 ps to equilibrate the system at a desired temperature, we began sampling the molecular positions and momenta at a uniform interval (1.0 ps) until the simulation time reached 100 ps. Figure 18 shows the 120,000 atoms of PE particle with a chain length of 100 beads.

5.1.5
Modeling of Bulk System

Fig. 18 also shows the PE bulk system in a cubic cell at the unit cell dimension of 4.6 nm. To create a polyethylene bulk system, a system consisting of 32 CH_2 chains with 100 C and 202 H atoms was placed in a cell, and the initial configuration of the system was optimized using a molecular mechanics calculation [230]. With the initial conformation, MD simulations were propagated by applying an external force to the cell every two fs, packing the chains into the cell with the desired density. The density was calculated using a cubic box with a dimension of 32 Å. The position of the box was randomly chosen in the cell, and then the number of the atoms in the box were determined. The process was repeated 10 times and the averaged density was monitored. The simulation was terminated when the density reached 0.85 g/cm^3 (bulk amorphous PE density).

We employed three-dimensional periodic boundary conditions to mimic the presence of an infinite bulk. The cell [Fig. 18(b)] was surrounded by the identical cells, allowing the atoms in the original cell to interact with all other atoms in the surrounding cells. Under the boundary conditions, classical trajectories were propagated for 10 ps. It is noted that we considered all atoms of PE polymer for the simulations of the bulk system in order to compare the computed structural

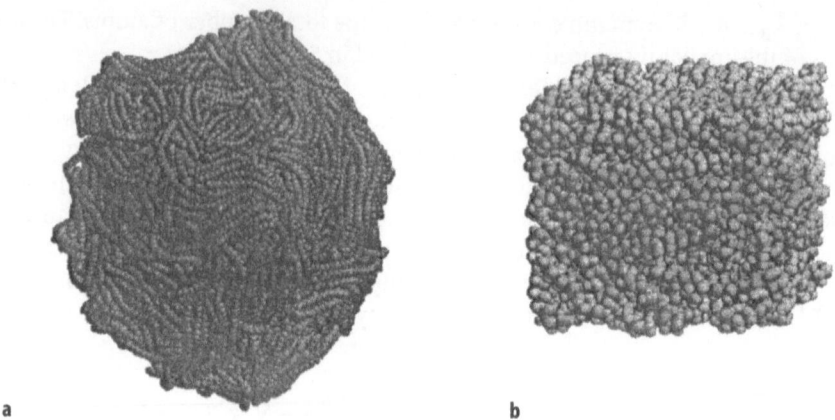

a   b

Fig. 18. a Polyethylene particle for 60,000 atoms with a chain length of 100 beads. The diameter for 60,000 atoms is 12.5 nm. The diameter of the particle is averaged values of distance from a center of mass to the surface atoms. **b** Polyethylene bulk system has 9664 atoms with chain length of 100 CH_2 with hydrogen atoms in the length of 4.6 nm cubic cell

conformations (e.g. radial distribution) with X-ray and neutron scattering experiments. The radial distribution of the simulated bulk PE was in good agreement with the experimental data. Comparing the bulk system with the nanoscale spherical particles, we studied the conformational changes of the particles due to the size reduction and the shape.

5.1.6
Surface Analysis

In analyzing surface effects of ultra fine polymer particles, we calculated the surface area and volume using Connolly's contact-reentrant method [231–233]. The contact-reentrant surface (molecular surface) uses a rigid hard sphere with a spherical probe that can trace molecular shape while in contact with the van der Waals surface. The van der Waals surface treats a CH_2 bead as a rigid hard sphere equal to the van der Waals radii. The union of the spheres is considered the volume. Connolly's method provides a smooth surface by patching the space between the probe radius and the probed molecule. For our surface area and volume calculations, the van der Waals radius for the PE particles is set at 1.89 C for CH_2 beads and the probe radius at 1.4 C to mimic the H_2O molecule. Using the molecular surface, we also computed the fractal dimension D to determine roughness of the surface.. The value of D can be calculated by

$$D = 2 - \frac{d \log A_s}{d \log R_p}$$

(20)

where A_s and R_p are the molecular surface area and probe radius, respectively [231]. The fractal dimension is determined by the slope and has values in the range 2–3; the lower limit corresponds to a smooth surface and the upper limits to a space-filled surface.

5.1.7
Thermal Properties

Experimental measurements of a melting point (T_m) and glass transition temperature (T_g) are crucial properties that can be used to assess the physical state of polymer [234]. In our simulations, the melting point and the glass transition temperature were obtained by calculating total energy and molecular volume of a system as a function of temperature [235]. For a transition from the amorphous (solid) to the melt phase (liquid) and a glass-rubber transition, the volume increased due to the conformational disorder of the polymer particles. Energy, temperature and volume were calculated while gradually annealing the system.. Computing the straight lines of molecular volume and total energy *vs.* temperature with a least squares fit, we took the points where those extrapolated straight lines met as the transition temperature T_m and T_g. In the annealing process, we controlled temperature by scaling the momenta in Eq. (18). The scaling parameter can be written

$$p_{i'} = \zeta p_i \qquad (21)$$

where p_i is the Cartesian momenta of the ith atom. In practice, the constant scaling parameter is set in the range from 0.90 to 0.995 to examine dependence of those transition temperatures and slope on the annealing rate. We propagated the classical trajectories with the initial configurations of the steady state of the amorphous PE particle at a temperature of 450 K and scheduled sampling the molecular positions and momenta at a uniform interval (1.0 ps) until the temperature reached 10 K.

5.1.8
Mechanical Property: Compressive (Bulk) Modulus

In order to investigate the deformation properties and mechanical memory, it is of interest to study the behavior of polymer fine particles while in an applied external force. One of the experimentally obtained values of those properties is known as Young's modulus, which is a fundamental measure of the stiffness of a material. Numerous calculations have been performed for the mechanical property of bulk-like crystalline PE polymer using force field [187–190] semi-empirical [236], *ab initio* calculation [237,238], and *ab initio* MD methods [239,240]., We have calculated the compressive (bulk) modulus for the amorphous PE particles using MD with an external force as shown in Fig. 19. At the start, a plate treated as a continuous wall is set at 10 Å from the closest atom of the particle in

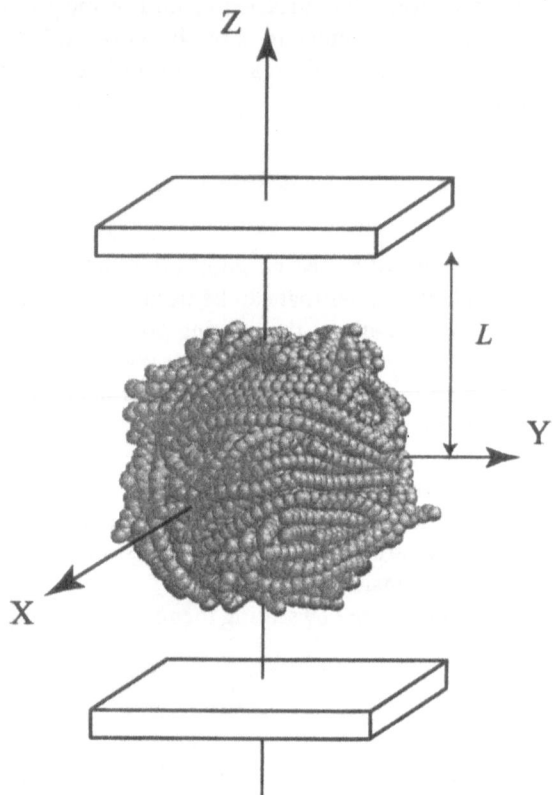

Fig. 19. Schematic of the geometry for stress-strain MD simulations

z coordinate (the distance from the origin to the plate is defined as L and the other plate was set at $-L$ in Cartesian coordinate.). The initial values of the Cartesian momenta were chosen randomly in the amorphous particle to set initial temperature less than the melting point (200 K) and the system was fixed with zero center-of-mass position and velocity. The plates were moved 0.1 Å every 0.5 ps and the total energy for L was calculated every 0.5 ps. Young's modulus is defined,

$$\sigma = Y\varepsilon \tag{22}$$

where and represent the tensile stress and strain, respectively. In our calculation, the strain is the fractional change in dimension, $=(L_0-L)/L_0$, and the stress is a force per unit area, $=F/area$. Since the external force is applied to the system, the particle starts deforming, and the total energy is fluctuated. We define L_0, which gives the minimum total energy of the system. To find the *area* (pr_{ave}^2) of the particles, the particles are projected on the xy plane, and the distance of each atom from the origin is calculated at the L_0. The distance r_{ave} can be approximately calculated from,

$$r_{ave} = \frac{\sum_{i=1}^{N}\left(x_i^2 + y_i^2\right)^{1/2}}{N} \tag{23}$$

The force applied to the particle is calculated from,

$$F = \frac{dE_T}{dL} \tag{24}$$

where E_T is the total energy of the system. By monitoring stress and strain, we obtained the compressive (bulk) modulus and yield point for the PE particles.

5.1.9
Collision of Polymer Particles

Classical trajectory calculations are used to develop an understanding of the experimentally observed behavior of collisions and surface interactions of nano- and micro-scale polymer particles on Si, C, and Al surfaces in a vacuum. The PE particles were propagated up to 100 ps for those surfaces to investigate the self-organization of polymer particles (mechanical memory) via collisions. At the start of classical trajectory, the initial values of the Cartesian momenta, p, were given randomly in the amorphous particle to set the initial temperature to 300 K, and the system was subjected to zero center-of-mass position and velocity. After the surface was set at 10 C from the closest atom of the particle in z coordinate, a desired initial translational velocity, $v_{z,init}$, was assigned to all atoms of the particle in z direction

$$p'_{z_i} = p_{z_i} + mv_{z,init} \tag{25}$$

where p_z is the initial momenta for the trajectories at time=0. Evaluating the non-bonded, long-range force on every atom and the surface every 0.02 ps, classical trajectories are propagated for 100 ps and the configurations and momentum are saved every picosecond. Figure 20 illustrates the collision of the PE particle of 12,000 atoms with a chain length of 100 beads on a Si surface at $v_{z,init} = 8$ Å/ps.

The instantaneous temperature of the particle is calculated from,

$$\frac{3}{2}Nk_B\langle T\rangle = \left\langle \sum_{i=1}^{N}\frac{p_i^2}{2m}\right\rangle - E_{trans}, \tag{26}$$

where k_B is the Boltzmann constant, N is the total number of atoms, and E_{trans} is the translational energy that can be obtained from the center-of-mass velocity of the particle. In order to investigate the overall particle shape, the interia ten-

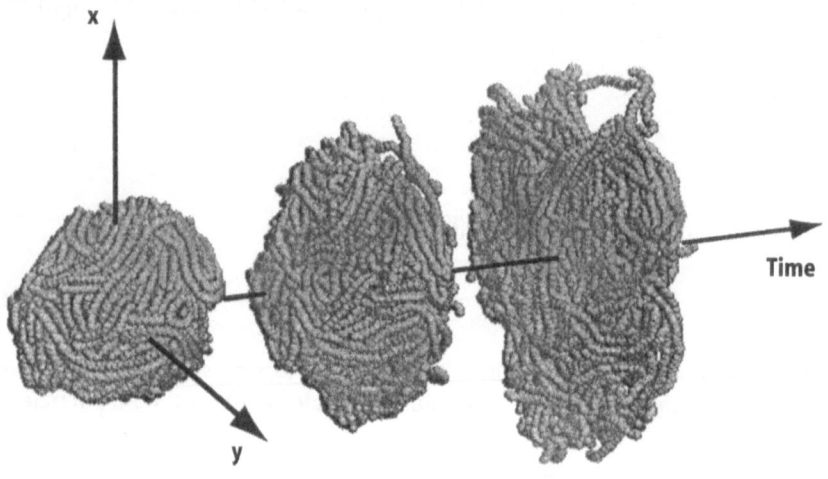

Fig. 20. The panels show evolution of the classical trajectories for the PE 12000 atoms with a chain length of 100 beads at time: t=0, 5.0 and 10 ps, respectively. The initial velocity, $v_{init,z}$ =, is set at 8 Å/ps

sor and the principle moments of inertia were calculated at picosecond interval during the simulation. An asymmetry parameter, κ, calculated from the principal moments of inertia provides the essential information for the shape

$$\kappa = \frac{2B-(A+C)}{A-C} \, , \tag{27}$$

where A, B, and C are inversely proportional to the corresponding moments of inertia ($A>B>C$). ranges from –1 for a prolate top, to +1 for an oblate top.

5.2
Characteristics of Nano-Scale Polymer Particles

Using the molecular dynamics technique, PE particles with chain lengths of 100 beads generated with up to 120,000 atoms have almost a spherical shape (asymmetry parameter –0.1 at a temperature of 10 K) as shown in Fig. 18, in good agreement with our experimental results. To interpret properties of the polymer fine particles differing from their bulk solid phase, we first counted the surface atoms using a three-dimensional grid method in Cartesian coordinates and the ratio of the surface atoms to the total number of atoms were obtained (the diameter is the average value of distance from a center of mass to the surface atoms). Since the smaller sized particles have a higher percent of atoms on the surface than the larger ones, a decrease in diameter increases the ratio as shown in Table 6. The large ratio of surface atoms to the total number of atoms causes a

Table 6. Surface effects in polymer particles

No. of backbone atoms	Diameter (nm)	Ratio of surface atoms	Area (nm^2)	Volume (nm^3)	S_{ratio} (nm^{-1})	Fractal Dimension (nm^{-1})
Polyethylene (PE)						
3000	4.7	72%	152	67	2.27	2.14
6000	5.6	70%	257	138	1.86	2.15
9000	6.4	69%	346	209	1.65	2.16
12000	7.0	67%	443	279	1.55	2.16
18000	8.1	65%	610	419	1.46	2.19
24000	9.0	64%	814	564	1.44	2.21
30000	9.7	62%	955	706	1.35	2.21
60000	12.4	60%	1879	1424	1.32	2.26
90000	14.1	56%	2765	2127	1.30	2.40
120000	16.1	54%	4035	3152	1.28	2.40
Polyethylpropylene (PEP)						
3000	4.9	70%	190	86	2.22	2.19
6000	6.1	67%	327	175	1.87	2.21
9000	7.0	66%	478	264	1.81	2.22
12000	7.7	63%	568	354	1.60	2.25
Polypropylene (PP)						
3000	5.3	66%	235	105	2.24	2.21
6000	6.8	65%	416	214	1.94	2.21
9000	7.3	61%	604	322	1.88	2.24
12000	7.9	60%	780	431	1.81	2.28
Polyisobutylene (PIB)						
3000	6.3	59%	332	131	2.53	2.25
6000	7.2	57%	683	262	2.60	2.28
9000	8.2	53%	948	399	2.38	2.32
12000	9.1	51%	1193	539	2.21	2.45

reduction of the non-bonded interactions for the surface layer; hence the cohesive energy is dramatically dependent on the size.

The surface area and volume are calculated using the contact-reentrant surface method with a probe radii, R_p=1.4 Å. The large proportional surface area defined by S_{ratio}=(surface area)/(volume) leads to large surface free energy, which is described by per unit of surface area (J/nm^2). The S_{ratio} of the particles is large compared to the bulk so that the surface area and surface free energy are large. The surface of the particles is also characterized by the fractal dimension, which describes a degree of irregularity of a surface [220, 227, 228]. Figure 21 shows dependence of surface area on probe radii for the PE particle of 30,000 atoms with a chain length of 100 beads. The slope is calculated for the probes in the range of 2.0 and 4.0 Å. The values of D are smaller for the particles than the value of the bulk (D=2.72). This indicates that the surface is irregular and has many cavities that might introduce unique (catalytic or interpenetrating) properties of polymer fine particles. Although a smaller probe can provide more detail of the polymer surface than a larger one, the fractal dimension for the probe in the range of 1.4 and 2.0 is physically unrealistic. As the size of the probe decreases, the area of the surface rapidly increases. Since the smaller probe penetrates the cavities on the surface, the values of the area and surface are strongly dependent on the probe. This predicts that nano-scale polymer particles are loosely packed and can show dynamical flexibility (e.g., compressive modulus of the particles is much smaller than that of the bulk system). The free volume (cavities) and molecular packing can be important in a diffusion rate of a small molecule trapped in the particles [230]. The structural characteristics of the PE particles up to 60,000 atoms with a chain length of 100 beads are shown in Table 6.

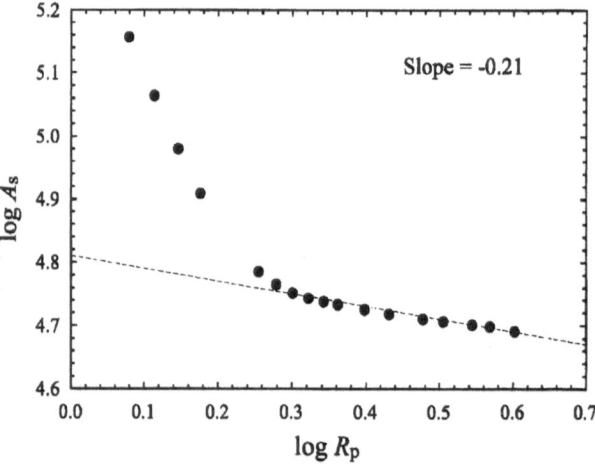

Fig. 21. Dependence of molecular surface area, A_s, on probes with radii, R_p, of 1.4 to 4 Å for polyethylene particle of 30000 atoms with chain length of 100 beads. The slope of straight line for log A_s vs. log R_p with the least squares fit is 0.21. The value of fractal dimension is evaluated for probe radii in the range of 2.0 to 4.0 Å.

5.2.1
Conformations

Figure 22 shows radial distribution functions for the PE particles of 12,000, 30,000, and 60,000 atoms with a chain length of 100 beads at a temperature of 100 K. For the amorphous PE particles, the peak positions of the radial distributions are insensitive to the size in the diameter range 16.1 nm. By comparing the peak positions of the radial distributions of the particles with the bulk system, it is clear that the peak around 3.15 Å corresponding to *gauche* configuration is very small for the particles. The reduction of *gauche* configuration in the radial distributions is believed to be due to alignments of the chains on the surface. In our previous study, we monitored the averaged end-to-end distances of the surface- and inner-chains for the particle of 12,000 atoms with a chain length of 100 beads. The average end-to-end distance of the surface chains is longer than that of the inner chains, and the inner chains have more gauche configurations than the surface [214].

Several simulations have been applied to study the morphology of single or multiple chains with different chain lengths. Since the surface chains of the PE nano-particles tend to straighten and align at temperatures below the melting point, the preferential morphology for the small PE particle with a long chain length is a rod-like shape. Liu and Muthukumar also observed this mechanism in the simulations of polymer crystallization. This stretching of the chains leads

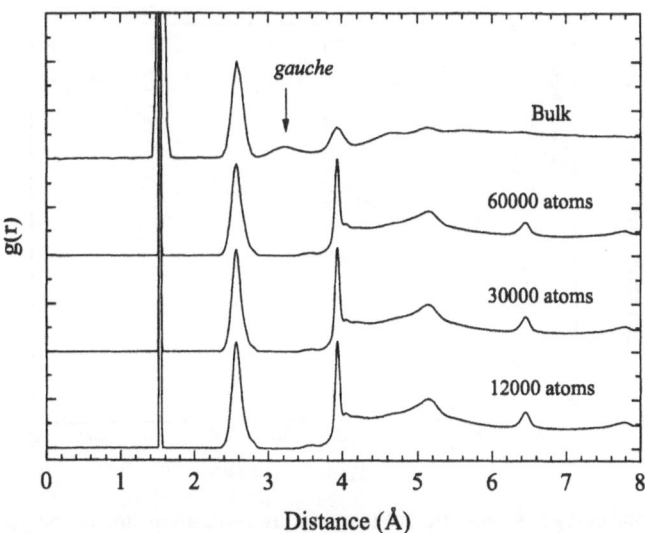

Fig. 22. Dependence of the radial distributions on size of polyethylene particles with a chain length of 100 beads. Those distributions are calculated from NHC constant temperature MD simulations. The three peaks are 1.54, 2.57 and 3.94 Å corresponding to the distance of $C_i - C_{i+1}$, $C_i - C_{i+2}$ and $C_i - C_{i+3}$ (*trans*), respectively.

to a reduction of the cohesive energy and an increase in volume. Studies on the effects of chain length show that the particles with the shortest chain length (50 beads) have the most spherical shape.

5.2.2
Thermal Properties

Thermal analysis has provided a great deal of practical, important information about the molecular and material world relating to equations of state, critical points, and the other thermodynamics quantities. For our study of thermodynamic properties of nano-scale polymer particles, we calculated temperature, volume, and total energy in the process of annealing the system by scaling the momenta. To investigate dependence of transition temperatures on annealing rates in our MD simulations, we examined the melting point and glass transition temperature for the particle of 12,000 atoms with a chain length of 100 beads with the annealing range from 1.7 to 29.5 K/ps. The transition temperatures were found to be rather insensitive (within the error of 5 K) to annealing rates slower than 6.0 K/ps. In all subsequent simulations, we set the annealing rate at approximately 2.5 K/ps (a constant scaling factor, ξ =0.992 in Eq. [21]) so that the particle was thermally equilibrated for each sampling point. Computing the straight lines of total energy vs temperature with a least square fit, we took the point where the extrapolated straight lines met as the melting point, T_m. Figure 23 shows the dependence of total energy of the system on temperature for the PE

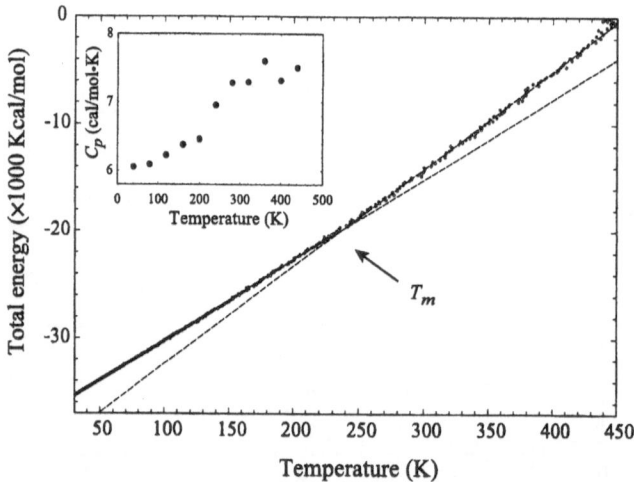

Fig. 23. The total energy as a function of temperature while annealing the polyethylene particle which consists of 12,000 atoms with a chain length of 100 beads. The *open circles* are the total energy calculated from MD simulations and the *dashed lines* show the least squares fit. The intersection of extrapolated straight lines corresponds to the melting point, T_m= 245 K. The slopes of the lines C_p are 6.47 and 7.24 cal/(mol·K), respectively. The insert shows the heat capacity calculated from the total energy

particle of 12,000 atoms with a chain length of 100 beads. The slopes provide the constant pressure heat capacity, C_p, above and below the melting point. Figure 24 shows the dependence of molecular volume for the PE particle as a function of temperature. From this volume calculation, we extracted the melting point and glass transition temperature. The melting points calculated from molecular volume for the different size of the PE particles with various chain lengths are close to those values calculated from total energy within an error of 10 K. The agreement in the melting points supports our hypothesis that it is possible to extract the melting point for a variety of nano-scale polymer particles from our MD simulations. We have reported the melting point from the simulation of energy *vs.* temperature instead of volume *vs.* temperature, since it is more straightforward to compute the linear lines from energy *vs.* temperature than from volume *vs.* temperature. Table 7 and 8 summarize thermal properties of the polymer particles with respect to size and various chain lengths.

Figure 25 shows the dependence of melting point and glass transition temperature on the diameter of the particles. The dramatic reduction of the melting point for the fine polymer particles is an example of surface effects and shows the importance of size. Since the large ratio of surface atoms to the total number leads to a significant reduction of the non-bonded interactions, the melting point decreases with the decrease of the total number of atoms. Figure 26 shows the effect of chain length on transition temperatures. A strong dependence of the melting point and the glass transition temperature on chain length is attributed to molecular weight and nonbonded energy of each chain. Since the large ratio of surface atoms to the total

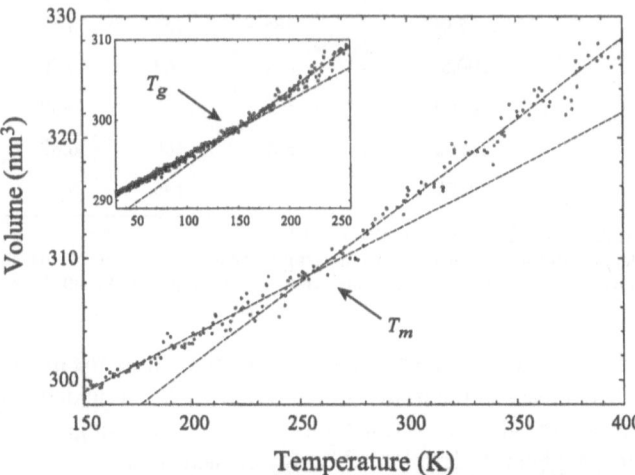

Fig. 24. The molecular volume as a function of temperature while annealing the polyethylene particle which consists of 12,000 atoms with a chain length of 100 beads. The open circles are the volume calculated from MD simulations and the dashed lines show the least squares fit. The intersection of extrapolated straight lines corresponds to the melting point, T_m=255 K. The insert shows the glass transition temperature T_g=157 K

Table 7. Thermal properties of polyethylene particles

No. of backbone atoms	Cohesive energy[a] (Kcal/mol)	T_m (K)	T_g (K)	C_p (cal/mol $\cdot$ K)	
				$T \leq T_m$	$T \geq T_m$
Chain length of 100 beads					
3000	3010	218	111	6.37	7.57
6000	6510	234	134	6.50	7.70
9000	10170	244	131	6.44	7.22
12000	13900	242	157	6.47	7.24
18000	23900	249	155	6.44	7.66
24000	32000	254	154	6.45	7.95
30000	41170	258	152	6.40	7.53
60000	82950	266	161	6.65	7.80
90000	125240	272	162	6.96	8.25
120000	166210	285	170	6.64	8.01
Chain length of 50 beads					
6000	7510	186	73	6.20	7.23
12000	15300	218	110	6.37	7.74
18000	25200	258	131	6.50	7.93
24000	34500	272	151	6.54	8.14
Chain length of 200 beads					
6000	5738	285	152	6.37	8.02
12000	10020	317	162	6.76	7.73
18000	18100	330	175	6.80	8.18
24000	25300	351	180	6.89	8.15
Bulk[b]	9073	414	195	5.19	7.67

[a] The values are calculated from NHC simulations at 10 K
[b] Cubic boundary conditions were used with a box length of 4.6 nm. The value of surface tension is for linear polyethylene at 20 K [Ref. 199]. The values of C_p are (from Ref. 204) at 300 K and 400 K

number leads to a reduction of the non-bonded interactions, the melting point decreases with a decrease of the total number of atoms. A strong dependence of the melting point and the glass transition temperature on chain lengths is attributed to the molecular weight and non-bonded energy of each chain.

To investigate the atomistic mechanism of melting, the melting for the surface and inner chains of the particles were investigated as a function of size, chain length, and temperature by calculating molecular diffusion near the melting point for PE nano-particles. We selected the surface and inner chains for the PE particles, using a three-dimensional grid method in Cartesian coordinates to

Table 8. Thermal properties of PEP, PP and PIB particles

No. of backbone atoms	Cohesive energy[a] (Kcal/mol)	Tm (K)	Tg (K)	Cp (cal/mol · K) T ≤ Tm	T ≥ Tm
Polyethylpropylene (PEP)					
3000	3279	218	110	7.80	9.23
6000	7014	220	120	7.87	8.89
9000	11160	238	132	7.91	9.09
12000	14520	242	140	7.99	8.78
Polypropylene (PP)					
3000	3500	225	115	6.42	6.70
6000	7350	233	127	6.33	6.86
9000	12920	241	140	6.42	6.80
12000	16230	248	146	6.50	6.76
Polyisobutylene (PIB)					
3000	4080	230	130	6.21	6.65
6000	7880	235	141	6.20	6.51
9000	12780	250	155	6.21	6.77
12000	18780	258	160	6.24	6.80

[a] The values are calculated from NHC simulations at 10 K

count surface atoms. When more than 80% of surface atoms were involved in a chain, the chain was defined as a surface chain; a chain that had less than 80% of surface atoms was considered an inner chain. The number of the surface and inner chains obtained from this procedure is 22 and 38 for the PE particle of 6000 atoms with a chain length of 100 beads.

Trajectories of center-of-mass for the surface and inner chains were calculated at every picosecond interval to analyze the diffusion coefficient at a temperature. Diffusion coefficients in the surface and inner chains were computed from mean-square displacement (MSD) [194,209,210]. The average mean-square displacement of center-of-mass for the chains can be determined,

$$< r^2(t) >= \frac{1}{N_{chn}} \sum_{i=1}^{N_{chn}} < \left| r_{CMi}(t) - r_{CMi}(t_0) \right|^2 > \tag{28}$$

where r_{CMi} is the Cartesian center-of-mass of ith chain and N_{chn} is the number of chains. The diffusion constant was calculated from this curve using the Einstein relationship by taking the value of the slope,

$$D = \frac{1}{6} \lim_{t=\text{infinity}} \frac{d}{dt} < r^2(t) > \tag{29}$$

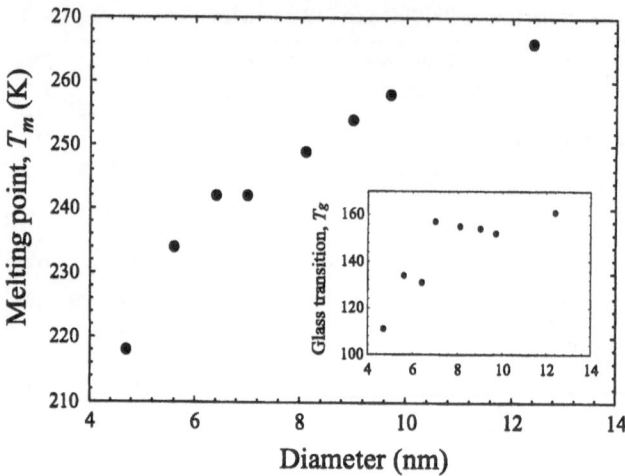

Fig. 25. Dependence of the melting points on diameter for PE particles with a chain length of 100 beads

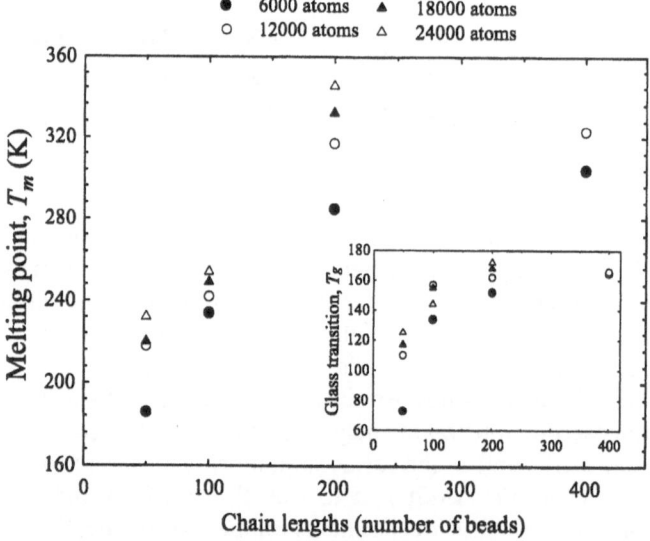

Fig. 26. Dependence of the melting points on chain lengths

Figure 27 shows the dependence of the relative diffusion on temperature for the surface and inner chains of 6000 atoms with chain lengths of 100 and 50 beads. It is noted that we report the relative diffusion D_{rel} by setting the diffusion coefficient of the inner chains of the PE particle of 6000 atoms with a chain length of 100 beads at a temperature of 100 K as one. The value of D_{rel} for the surface chains is larger than one of the inner chains. For the low temperature ($200K \leq T$),

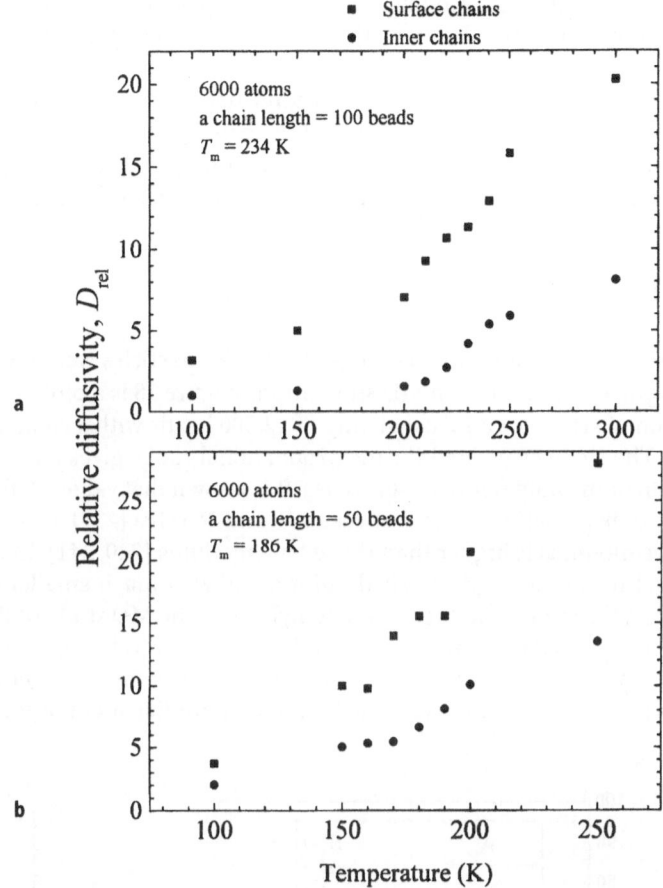

Fig. 27. Relative diffusion constant of the surface and inner chains in the PE particle of 6000 atoms with chain length of **a** 100 beads and **b** 50 beads. The *closed circles* and *triangles* show the diffusion for the surface and inner chains of the particle

the diffusion of the particle with a chain length of 100 beads was not significantly changed much. For the temperature above 200 K, however, the value for the surface chains rapidly increased. The higher diffusivity of the surface chains provides evidence that the melting starts from the surface of the PE particles, and the value of D_{rel} increases as molecular weight decreases. From our thermodynamic analysis, the melting point of the particle (6000 atoms) with chain lengths of 50 and 100 beads was found to be 186 and 232 K, respectively. The melting point was higher than the temperature at which the surface chains experienced a significant change in mobility.

The plot for the surface chains gives two steps of the melting transition. The first step (mechanical melting) is at a temperature of 200 K and the second (thermodynamic melting) is 230 K. The first step is an early melting stage at which all chains in the particle starts moving. The second step is the fusion stage of a solid. There is a plateau region between the first and second stages. The mobility of the chains in the outer layers increases rapidly after the region. In contrast, the chains in the core layer do not have the second step and increase linearly. This observation provides the surface melting of the nano-scale particle. The melting point obtained from the total energy vs temperature plot has good agreement with the temperature of the second stage of the diffusion coefficients.

5.2.3
Mechanical Properties

The compressive modulus and yield point for the PE particles were investigated by applying an external force in MD simulations. Figure 28 is a plot of the stress as a function of strain for a PE consisting of 12,000 beads with a chain length of 100 beads. The slope calculated for the proportional range gives a modulus for compression of the polymer nano-particles. It is known that values of the tensile modulus of bulk polyethylene are between 210 and 340 GPa [239]. In general, the compressive modulus is higher than the tensile modulus [240, 241]. In addition, the bulk and tensile strength or yield point are always much smaller than the modulus for a thermoplastic such as polyethylene. In the MD study of the nanoparticles, we observed a compressive modulus that is several magnitudes of order smaller than the bulk values, and the yield point is much larger than the modulus. The stress-strain curve actually looks more like a curve for an elas-

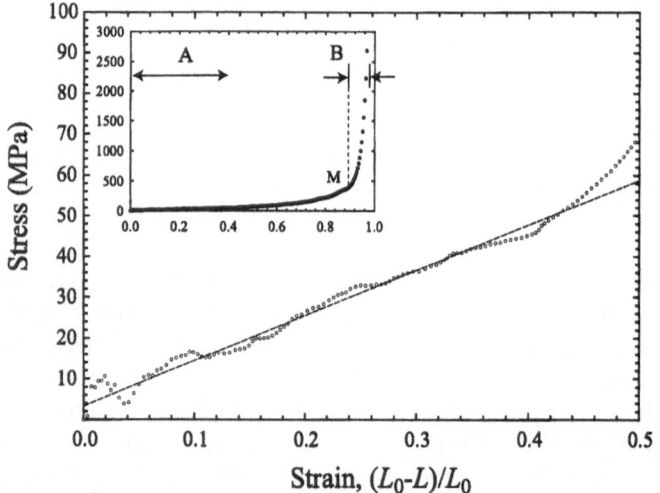

Fig. 28. Stress-strain behavior for the PE particle of 12,000 atoms with a chain length of 100 beads. The yield point is indicated at point M. The region of non-reversible deformation caused by the compression is labeled A; the region of reversible deformation is labeled B. The slopes calculated in the strain range from 0 to 0.4.(region A) and from 0.86 to 0.97 (region B) are 111 MPa and 85 GPa, respectively

tomer. However, the initial deformation caused by the compression (that which gives the modulus) is not reversible (region A in Fig. 28). What occurs during this phase is the deformation of a spherical particle to an oblate top as shown in Fig. 29. This structure is stable but it lies at a slightly higher energy than that for a spherical particle. Thus, the modulus for compression in this region is actually more a measure of the force required to deform the spherical polymer particle into an oblate top (pancake-like structure). Further deformation tends to be reversible up to the point of rupture (region B in Fig. 28). This deformation is actually more closely related to the bulk modulus since the stress is due to the cohesive energy and not a microstructure. This leads to a yield point that is significantly larger than the modulus. Table 9 shows the dependence of mechanical properties of PE particles on size.

Table 9. Mechanical properties in ultra fine PE particles[a]

No. of atoms	Compressive modulus (A) (MPa)	Compressive strength[b] (MPa)	Compressive modulus (B) (GPa)	Yield length[c] (Å)
3000	68	288	46	18.3
6000	86	324	63	33.6
9000	88	327	67	38.9
12000	111	400	85	40.0

[a] The modulus is calculated from the region of non-reversible and reversible deformation. (See Fig. 28, label A and B)
[b] The strength is obtained at the point, M
[c] The length is the thickness of the PE particle in z coordinate at the point, M

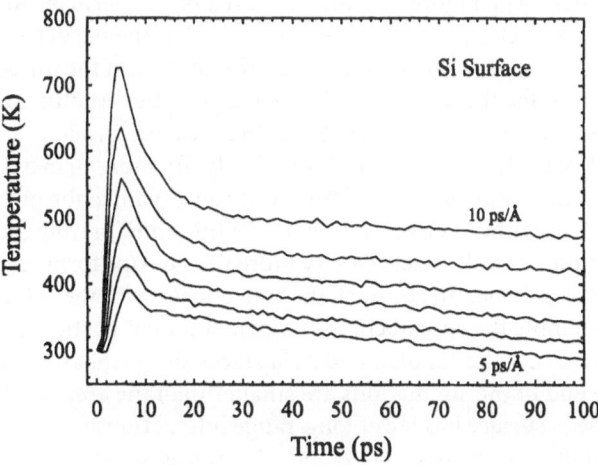

Fig. 29. Dependence of the temperature for the PE 12,000 atoms on initial velocities as a function of time. The initial temperature is set at 300 K. The solid lines show the temperature at $v_{init,z}$=5, 6, 7, 8, 9, and 10 Å/ps

5.2.4
Collision Dynamics

MD simulations were analyzed to elucidate the behavior of collisions and self-organization resulting from the mechanical memory of the nano-scale PE particles on three different surfaces, Si, C and Al. To probe the energy transfer via collision on the surface, we first calculated the temperature of the particle as a function of time. For all simulations, the initial temperature was set at 300 K that is above the melting point of the PE particles in the size-range of our study. Figure 29 shows the temperature dependence of the 12,000 PE atoms on initial collision velocities, $v_{z,init}$, for the Si surface. Since the translational energy of the system transfers to vibrational energy during collision with the surface, the particle has its highest temperature on impact. As would be expected, an increase of the initial translational velocity causes a higher temperature on the surface due to conservation of total energy. After impact, the temperature starts decreasing and the system reaches equilibrium. In the equilibrating process, the vibrational energy transfers back to translational energy and inter- and intra-non-bonded energy to reorganize the configuration. We also investigated the temperature for C and Al surfaces, and the results are similar to the Si surface for the range of the initial velocities.

One way of investigating the shape and critical velocities of the particle is to calculate the area obtained by projecting the particle on the xy plane. The area is approximated from $area_{xy}=pr_{ave,xy}^2$ where $r_{ave,xy}$ is the averaged distance of the atoms from the center-of-mass on the plane. The critical velocities for the collision can be divided into two parts : 1) the lowest initial velocity that the PE atoms lose some chains via the collision, and 2) the highest initial velocity that the PE atoms can self-organize back into the spherical shape after the collision (mechanical memory). Figure 30 shows the area of the particles for several initial velocities on Si, C, and Al surfaces. It is clear that the particles with $v_{z,init}$=9 and 10 Å/ps are spread on the surface and have totally lost the initial memory of the conformation for those surfaces. To investigate the structural details of the collisions, we used three-dimensional graphical visualization. From this study, we found that particles with a critical initial velocity more than 8 Å/ps lost several chains via the collision. But polymer can self-organize for initial velocities lower than 7 Å/ps without losing any chains. While equilibrating the system, the particles with the critical initial velocity more than 8 Å/ps form several separated blocks of the polymer. In contrast, the trajectories with the velocity less than 7 Å/ps self-assemble the spread chains to form a united particle. A comparison of the collision dynamics for Si, C and Al surfaces shows that the areas for the Si surface at the end of the simulations are smaller than the areas of the other surfaces. Since the Si surface has fewer long-range interactions (swallower potential well) with the PE particle than the C and Al surfaces, the particle on the Si surface can reorganize the shape easier than that on the other surfaces. Figure 31 shows dependence of the area on the size for the initial velocities at 7 and 8 Å/ps. The area with the initial velocity at 7 Å/ps increases with an increase in the size

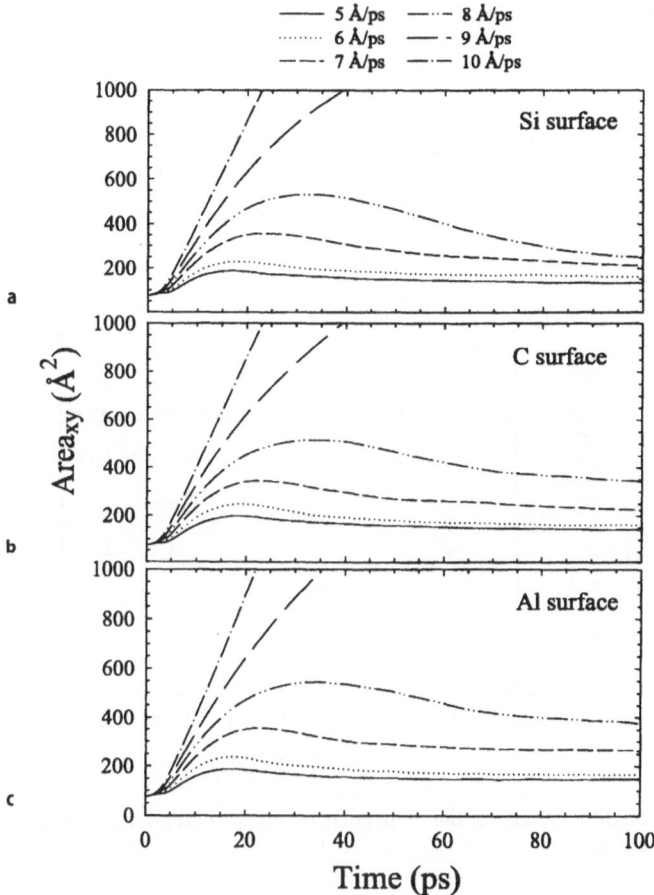

Fig. 30. Dependence of the area for the PE 12,000 atoms on initial velocities as a function of time for **a** Si, **b** C and **c** Al surfaces. The initial temperature is set at 300 K. The *lines* show the area at $v_{init,z}$=5, 6, 7, 8, 9, and 10 Å/ps

of the PE atoms, and the particles reassemble without losing any chains. The figure also shows the dependence of self-organization on the size at 8 Å/ps. For the initial velocity, the 3,000 and 6,000 PE atoms reform a united particle, but the 9,000 and 12,000 atoms do not reform. This result suggests that a smaller polymer particle can reform a united particle more efficiently than a larger one and the critical velocity can be faster.

We investigated the mechanistic memory of the PE 12,000 atoms to reform a sphere-like shape for the Si, C, and Al surfaces by calculating the asymmetric parameter in Eq. (20). As shown in Fig. 32, the particle with the initial velocity 5 Å/ps tended to self-organize into the spherical shape during equilibration. Since the particles spread (coats) on the surfaces on impact, the asymmetric parameter had the largest value at the impact (oblate top). It was interesting to note

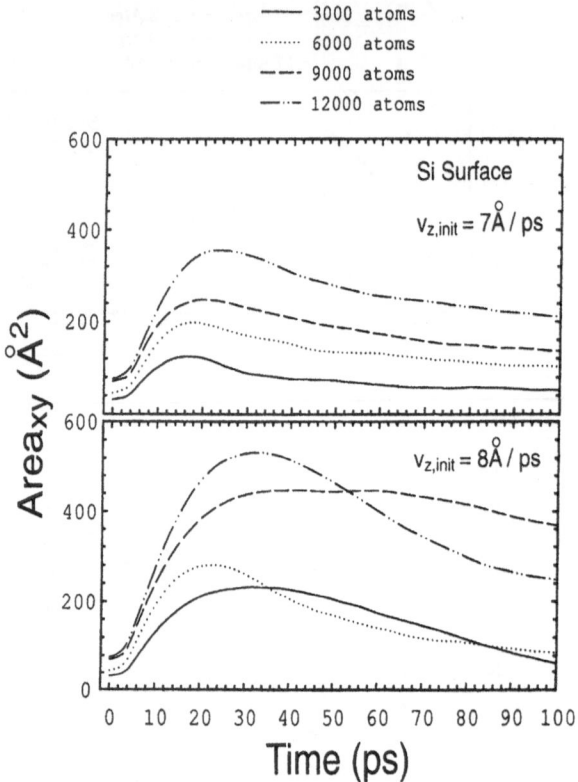

Fig. 31. Area on different size of the PE particles at $v_{init,z}$=7 and 8 C/ps. The initial temperature is set at 300 K

that most of the particles end up with the shape of a prolate top for those surfaces. For high velocities, the particles can not self-organize back to their original shape. This is due to the fact that the chains for the particles on impact are broadly spread on the surface and have the two competing interactions :the non-bonded chain-chain and the chain-surface interactions. In the process of self-assembling the chains for the high velocities, the temperature of the system starts decreasing and the chains of the particle tend to align to minimize the potential energy. As a result, the particles have a preferential shape of a prolate top [175, 212]. For those surfaces, the particles with initial velocities less than 5 Å/ps persist in the original spherical shape and have weak sensitivities to the collision dynamics on the different substrates. Figure 33 represents the shape of a particle for two limiting cases of an oblate and a prolate top for the Si surface at time= 60 ps. In the Fig. 33(a), the particle with the initial velocity 6 Å/ps has a pancake-like shape (oblate top : k=0.78). The preferential shape of a prolate top (k=−0.79) is shown in Fig. 33(b). In the figure, the particle with the initial velocity 8 Å/ps lost the mechanical memory while self-assembling the chains.

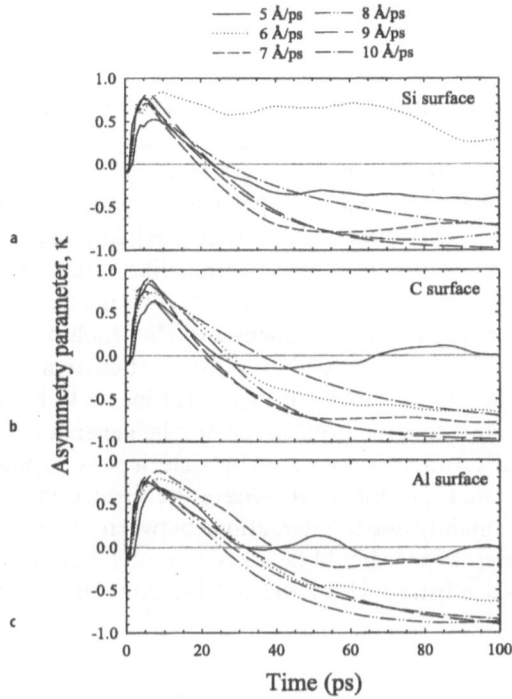

Fig. 32. Asymmetry parameter for the PE 12,000 atoms on initial velocities as a function of time for **a** Si, **b** C and **c** Al surfaces. The *gray lines* show a spherical shape

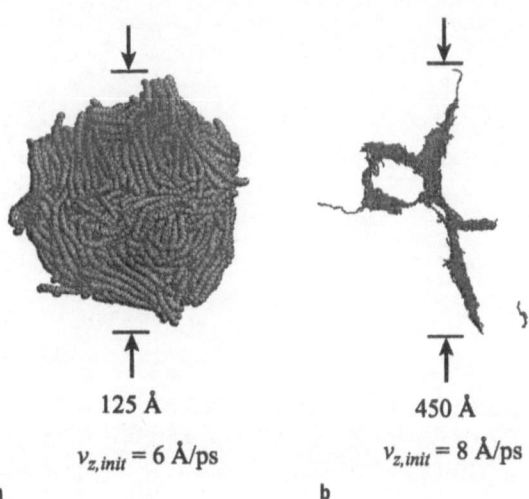

Fig. 33. Snapshots of the PE 12000 atoms for **a** initial velocity 6 Å/ps and **b** 8 Å/ps at time = 60 ps. The asymmetry parameters, k, are **a** 0.78 (oblate top) and **b** – 0.79 (prolate top), respectively

5.2.5
Surface Interaction

In the submicron liquid droplets experiment, polymer particles were deposited on gold or glass microscope slides, and their morphology analyzed by scanning electron, optical, or phase contrast microscopy [185,186]. To determine its size, shape, composition and phase separation, the particle was caught in an electrodynamic trap and analyzed using Franhofer diffraction [209]. In small sizes (1 m), the particles formed a mixture of spheres and "pancake-like" structures on a substrate as shown in Fig. 34 [Ref. 183]. The droplet generator was normally mounted vertically and the polymer droplets ejected downward. The whole assembly and the particle free-falling region were shielded from room air current disturbance. We believe that a typical terminal velocity of a particle (due to the bouncy of air) is several meters per second (~1/100 Å/ps) in the droplet generator.

Since a typical terminal velocity of particle is several meters per second (~1/100 Å/ps) in the experiment, we believe the distortion of a particle (pancake) structure is mainly due to interactions between the particle and the substrate, and the interactions should be a key to understanding polymer nucleation behavior on a substrate [196] and wetting phenomenon for a liquid on a

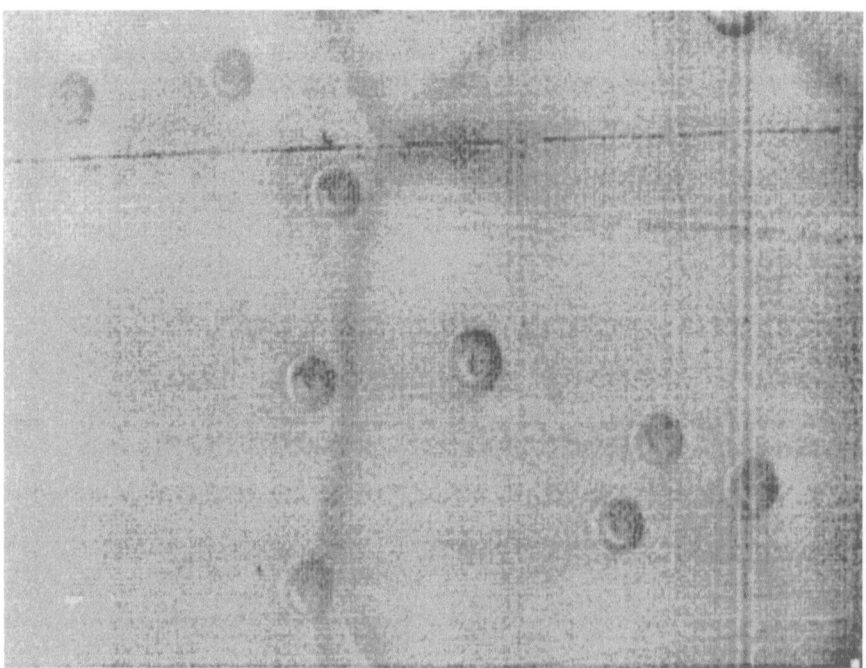

Fig. 34. Scanning Electron Microscope image of 3μm polyvinyl chloride particles on gold substrate. The shape of the particles is a 'pancake'-like structure

surface [189, 191, 213–215]. We used NHC molecular dynamics to explore the behavior of the PE particle touching the Al substrate at a temperature of 200 K, which is below the melting point. In the simulations, the surface was set at 5 Å from the closest atom of the particle in z coordinate, and the initial values of the Cartesian momenta were chosen with a random orientation in phase space and magnitudes such that the total kinetic energy is the equipartition theorem expectation value [216]. Figure 35 shows the area and asymmetry parameter of the PE particle of 3000 atoms with a chain length of 100 beads on a Al surface in comparison with NHC MD without the surface interaction. It is clear that the particle is attracted to the surface and the shape is distorted. The asymmetry parameter indicates that the particle ends up with the shape of a prolate top as shown in Fig. 35(b) (The surface chains of the particle of 3,000 atoms tend to straighten and align at temperatures below the melting point, hence the prefer-

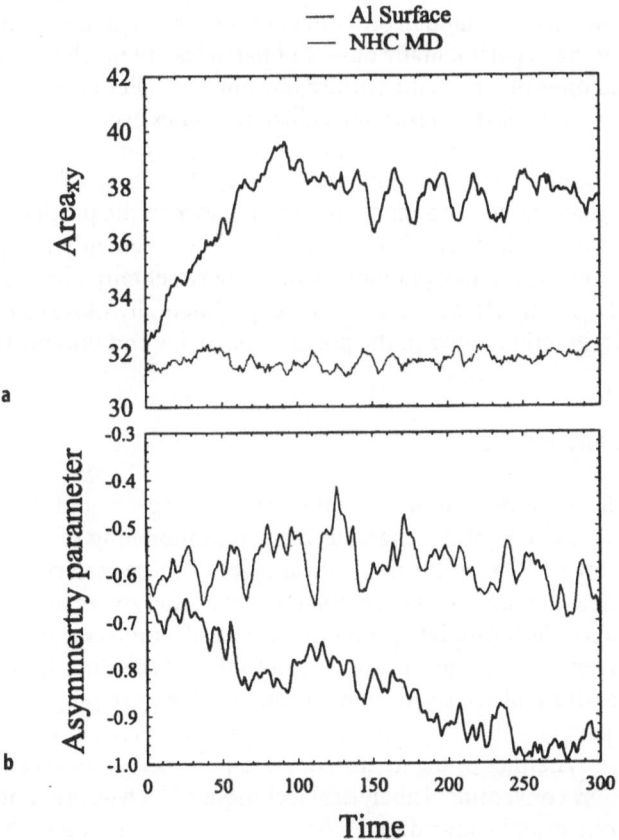

Fig. 35. Dependence of area and asymmetry parameter on time for the 3000 PE atoms. The area and asymmetry parameter were calculated for the particle with Al surface (*solid line*) and without the surface (*dashed line*) using the constant temperature MD

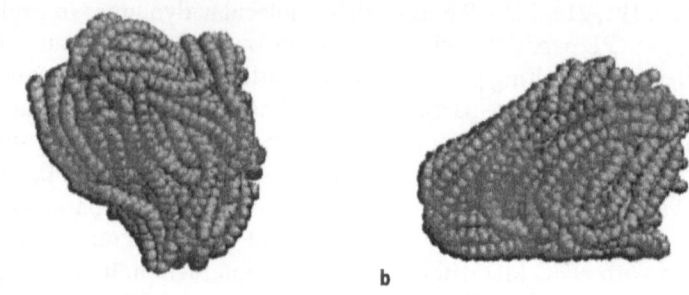

a   b

Fig. 36. Snapshots of the PE 3000 atoms at time **a** 0 ps and **b** 300 ps. The plates not shown represent the Al surface

ential shape is a prolate top [−0.6]). The initial and final structures are visualized in Fig. 36. The figure shows that the longest principle axis is parallel to the surface and the particle oriented in order to have the largest area in contact with the surface. Since the experimentally observed particles are much larger (1 m) than the simulated ones (4.7 nm), the surface tension is smaller and the distortion of the structure can be larger. From the collision studies, we found that the particles with the initial velocity less than 5 Å/ps self-organize into the spherical shape. In our collision simulations, the translational energy of the particles is much larger than the non-bonded interaction between the particle and the surface so that the particles can bounce back and subsequently reorganize their shape. Considering the morphology of the experimentally observed particles, we believe the main effect leading to the experimentally observed structure is the surface interaction between the polymer particles and the substrate.

6
Conclusions and Future

In this article, we have reviewed the most recent progress and discussed some new insights into generating, characterizing, and modeling of polymer micro- and nano-particles. One of the main advantages of these materials is the observation that new dynamic behavior emerges when polymers are confined in a small microparticle or droplet of solution the size of a molecule or molecular aggregates. Solvent evaporation in the droplet takes place on a time scale short enough to frustrate phase separation, producing dry pure polymer or polymer blend microparticles that are homogeneous within molecular dimensions. The structure and dynamics of the micro-confined polymer particle can be conveniently probed by conventional analytical techniques (such as Fraunhofer diffraction, fluorescence, and conventional phase-contrast microscopy). This capability allows production of useful polymer particles such as spherical polymer alloy microparticles with tunable properties as refractive index simply by adjusting the relative weight fractions of the polymers in solution. With the advent of advanced analytical tools and increasing interest in interdisciplinary research, it is likely that progress will continue to disclose additional different types of polymer particles and their unique possibilities.

Work to date has revealed that the GAP can be used to prepare powders within the size range of 0 to 200 µm from low-viscosity PE-based polymers and epoxy resins. Powders with properties tailored to varying applications can be efficiently prepared in short cycle times by changing relatively few process control variables in a contamination-free environment, making the GAP a useful alternative to conventional grinding processes. The sphere/fiber morphology of the atomized material is an advantageous combination for solid-state compaction due to the increased number of contact points as well as the self-reinforcing effect of the high aspect ratio microfibers. These benefits of the GAP process together with its flexibility, high throughput, and facile nature can be expected to make it highly attractive to industrial processes that must be safe, capable of mass production, and operate in an environmentally-benign manner. Investigations are needed into expanding the use of GAP to other polymers with vastly different thermal and rheological properties.

Statistical and computer neural network modeling can be applied to the GAP to obtain models that can be used to predict performance of the process in the forward and reverse modes. The neural network model predicts the performance to within an average of 12% for the forward and 35% for the reverse modes, making it a useful model for expanding the GAP technology to other systems without the need to perform large and expensive experiments.

The combination of experimental evidence and computational modeling shows conclusively that stable, homogeneously blended (bulk-immiscible) mixed-polymer composites can be formed in a single microparticle of variable size. To our knowledge, this represents a new method for suppressing phase-separation in polymer-blend systems without compatibilizers that allows formation of polymer composite micro- and nanoparticles with tunable properties such as dielectric constant. In order to form homogeneous particles, conditions of rapid solvent evaporation (e.g. small [<10 µm] droplets or high vapor pressure solvents) and low polymer mobility must be satisfied. While this work obviously focused on polymeric systems, it should be pointed out that the technique is easily adaptable to making particles of small organic and inorganic (and hybrid composites) as well. A wide range of electronic, optical, physical and mechanical properties of single- and multi-component polymer nanoparticles remain to be explored.

Currently, we are investigating structure and dynamics of a number of polymer-polymer, and polymer-inorganic composite systems. The latter, for example, shows an interesting time dependence in size, morphology, and phase-separation behavior whose underlying microscopic mechanism is not completely clear. In other work, we are exploring the role of compatibilizers (short-chain copolymers) to increase the size range for producing homogeneous polymer-blend particles. Also, we are developing techniques for multi-color diffraction to extract information on ternary and higher-order composite particles. Some important issues relevant to commercialization remain to be resolved as well. Most importantly is the problem of particle throughput that is currently (optimistically) limited to a few milligrams per day. Other issues include hardware compatibility with various solvents that are commonly encountered in polymer solution work. Low-vapor pressure solvents such as methylene chloride are seriously problematic in acoustically driven droplet ejection devices such as ours. In order to expand the practical range of materials amenable with this technique, effects such as cavitation and clogging due to solvent evaporation must be confronted.

To get deeper insights into the structure, dynamics, and properties of the polymer micro- and nano-particles, we have introduced an efficient method for modeling initial conditions for polymer nano-scale particles. Using molecular dynamics, we have simulated polyethylene nano-particles generated with up to 300,000 atoms with a variety of chain lengths. Our results show that the ratio of surface atoms to the total number of atoms for the polyethylene particles is very large, and the surface effects provide interesting properties that differ from the bulk system. In particular, the melting point and glass transition temperature were found to be dramatically dependent on the size of the polymer particles. The compressive modulus of the PE particles is significantly smaller than the value of the bulk, and the polymer particles act like an elastomer. It is well known that stress-strain properties are very sensitive to temperature, and an increase in temperature decreases a modulus. In the future study of mechanical properties of polymer nano- and micro-particles, we will analyze the dependence of the strength on size, temperature, chain length, and composition (multi-component blends). We are also extending this model to larger size particles and applying this method to model various polymer. As used in our work, molecular dynamics simulations can be expected to provide useful insights to interpret the properties and behavior of ultra-fine particles in new materials and devices. It is noteworthy that these particles can easily be electrically charged, and semi-classical studies of the electronic properties of these particles have been performed.

Finally, we draw attention to a problem somewhat neglected in this review: What information does the diffraction pattern carry with respect to size and morphology of submicrometer phase-separated structures within the polymer particle? That's the critical question surrounding the application of those techniques to polymer blend systems. The reason for this omission was that very little is known about this question to our knowledge. A joint experimental-theoretical modeling effort is needed to tackle this problem in the future, results of which will complement those from researchers working on traditional approaches. On the other hand, the subject of phase separations in polymer blends and block copolymer melts in thin film geometry has been recently reviewed by Binder [242] and Cheng and Keller [21] reviewed the role of metastable states in polymer phase transitions. We hope that the present review will provide useful guidelines to future experimental studies and theory development for the little-studied polymer and polymer blend micro- and nano-particles.

Acknowledgements. Partial support of JUO's research from the U.S. National Science Foundation (DMR9982077) and Huntsman Chemical Corporation, and the research work of his former graduate students are gratefully acknowledged. This research was sponsored by the U.S. Department of Energy, Office of Basic Energy Sciences (Divisions of Chemical Sciences and Materials Science), under contract DE-AC05–96OR22464 with Oak Ridge National Laboratory, and Laboratory-Directed Research and Development Seed Money Fund managed by Lockheed Martin Energy Research Corporation. We also acknowledge contributions from Professor K.C. Ng (ORNL Faculty Research Participation Program), K. Runge, and J.V. Ford (ORNL Postdoctoral Research Program). We thank Pam Reinig for a critical reading of the manuscript.

References

1. Narkis M, Rosenzweig N (eds) (1999) Polymer powder technology. Wiley, New York
2. Misev TA (1991) Powder coatings chemistry and technology. Wiley, New York
3. Ford WT (ed) (1986) Polymeric reagents and catalysts. ACS, Washington D.C.
4. Ford WT, et al. (1988) Pure & Appl Chem 60:395
5. Frechet JMJ et al. (1988) Pure & Appl Chem 60:353
6. Laszlo P (ed) (1987) Preparative chemistry using supported reagents. Academic Press, New York (and pertinent references contained therein)
7. Adachi C, Hibino S, Koyama T, Taniguchi Y (1997) J Appl Phys. 2:L827
8. Granstrom M, Inganas O (1996) Appl Phys Lett 68:147
9. Abraham D, Bharathi A, Subramanyam SV (1996) Polymer 37:5295
10. Otaigbe JU, Noid DW, Sumpter BG (1998) Advances in Polymer Technology 17:161
11. Otaigbe JU, McAvoy J (1998) Advances in Polymer Technology 17:145
12. Otaigbe JU, McAvoy J (1997) Materials World 5:383
13. Kung C-Y, Barnes MD, Sumpter BG, Noid DW, Otaigbe JU (1998) Polym Prepr Am Chem Soc Div Polym Chem 39:610
14. Nose T (1987) Phase Transitions 8:245
15. Hashimoto T (1988) Phase Transitions 12:47
16. Bansil R, Liao G (1997) Trends in Polymer Science 5:146
17. Binder K (1983) J Chem Phys 79:6387
18. Binder K (1990) Materials Science and Technology 5:405
19. Hair DW, Hobbie EK, Nakatani AI, Han CC (1992) Chem Phys 96:9133
20. Bates FS, Wiltzius P (1989) J Chem Phys 91:3258
21. Cheng SZD, Keller A (1998) Annu Rev Mater Sci 28:533
22. Bates FS, Fredrickson GH (1999) Physics Today, 32 (and references therein)
23. Hamley IW (1999) Block copolymers, Oxford Univ. Press, Oxford
24. Ito K (1998) Prog Polym Sci 23:581
25. Ito K, Seigou K (1999) Adv Polym Sci 142:129
26. Fukui K, Sumpter BG, Barnes MD, Noid DW, Otaigbe JU (1998) Polym Prepr (Am Chem Soc, Div Polym Chem) 39:612
27. Fukui K, Sumpter BG, Barnes MD, Noid DW, Otaigbe JU (1998) Proc 216th ACS National Meeting, Boston, (Book of Abstracts)
28. Kung C-Y, Barnes MD, Sumpter BG, Noid DW, Otaigbe J (1998) Polym Prepr (Am Chem Soc, Div Polym Chem) 39:610
29. Barnes, MD, Kung C-Y, Lermer N, Sumpter BG, Noid DW, Otaigbe JU (1999) Optics Lett 24:121
30. Scholz SM, Carrot G, Hilborn J, Dutta J, Valmalette J-C, Hofmann H, Luciani A (1998) Mater Res Soc Symp Proc 501:79
31. Deki S, Yano T, Kajinami A, Kanaji Y (1998) Plast Eng 43:95
32. Masui T, Machida K, Sakata T, Mori H, Adachi G (1997) J Alloys Compd 256:97
33. Noguchi T, Gotoh K, Yamaguchi Y, Deki S (1991) J Mater Sci Lett 10:477

34. Hifumi E, Kanmei R, Ishimaru M, Shimizu K, Ohtaki M, Uda T (1996) J Ferment Bioeng 82:417
35. Uda T, Kanmei R, Akasofu S, Hifumi E, Ohtaki M (1994) Anal Biochem 218:259
36. Tamai H, Sakurai H, Hirota Y, Nishiyama F, Yasuda H (1995) J Appl Polym Sci 56:441
37. Chen L, Yang W, Yang C (1996) Macromol Symp 105:235
38. Ohtaki M, Komiyama M, Hirai H, Toshima N (1991) Macromolecules 24:5567
39. Leslie-Pelecky DL, Zhang XQ, Rieke RD (1996) J Appl Phys 79(8, Pt. 2A):5312
40. El-Shall MS, Slack W (1995) Macromolecules 28:8456
41. Hayashi T, Mizuma K, Yao H, Takahara S (1993) Spec Publ – R Soc Chem 137:197
42. Chan HSO, Ming L, Chew CH, Ma L, Seow SH (1993) J Mater Chem 3:1109
43. Ohtaki M, Oshima, Y, Eguchi K, Arai H (1992) Chem Lett 11:2201
44. Tsubokawa N, Kogure A (1991) J Polym Sci Part A:Polym Chem 29:697
45. Neinhaus GU, Plachinda AS, Fischer M, Khromov VI, Parak F, Suzdalev IP, Gol'danskii VI (1990) Hyperfine Interact 56:1471
46. Otaigbe JU, McAvoy JM, Anderson IE, Ting J, Mi J, Terpstra R (1997) U.S. Patent Appl 08 895 645 (Filed July 17, 1997)
47. Handyside TM, Morgan AR (1992) GB Patent 2 246 571
48. Noid DW, Otaigbe JU, Barnes MD, Sumpter BG, Kung CY (1998) US Patent Appl 09 128 333 (Filed August 3, 1998)
49. Aoki K, Koide K, Matsumoto A (1996) Jpn Patent 08 157 532
50. Oda M, Setoguchi K (1990) Jpn Patent 02 097 501
51. Yamazaki M, Takebe K (1989) Jpn Patent 01 198 634
52. Shigemitsu M (1989) Jpn Patent 01 006 035
53. Kometani Y, Fumoto S, Tanigawa S (1972) Ger Patent 2 063 635
54. Akaishi M, Serizawa T, Taniguchi K (1998) Jpn Patent 10 337 794
55. Yamada H (1998) Jpn Patent 10 330 670
56. Suwabe Y, Mikami T, Fukuda M (1996) Jpn Patent 10 140 057
57. Uraki H, Sawada S, Iida Y, Fujigamori T, Hazama S (1997) Jpn Patent 09 053 035
58. Uraki H, Satake J, Iida Y, Fujigamori T, Hazama S (1997) Jpn Patent 09 053 036
59. Funaki Y, Tsutsumi K, Hashimoto T, Harada M (1998) Euro Patent 864 362
60. Ehrat R, Watrinet H (1997) Euro Patent 761 743
61. Tamura H Jpn Patent 04 025 102.
62. Nakai H, Edakawa S (1990) Jpn Patent 02 202 922
63. Kagawa A (1988) Jpn Patent 63 120 740
64. Kumaki J (1986) Euro Patent 197 461
65. Hayashi S, Murakami S, Goto K, Noguchi T, Yamaguchi Y (1995) Jpn Patent 07 025 642
66. Suda Y, Murakami M, Makino K (1994) Jpn Patent 06 287 313
67. Tamura H (1991) Jpn Patent 03 163 805
68. Kawazu K, Seike K, Kawamura T, Kiyata H, Onoyama H (1997) Jpn Patent 09 286 948
69. Katsushima A (1994) Jpn Patent 06 073 334
70. Fujiwa T, Urabe S (1997) Jpn Patent 09 255 914
71. Yoshihara T (1997) Jpn Patent 09 110 909
72. Kishimoto Z (1990) Jpn Patent 02 212 562
73. Morita H, Ishizaki Y, Azuma J (1989) Jpn Patent 01 170 677
74. Myanaga S, Doi Y, Tsunoda J (1996) Jpn Patent 08 259 846
75. Shichizawa A, Aachi, Ii-H (1995) Jpn Patent 07 138 304
76. Kitayama T, Takami N, Urabe K, Fukuda T (1994) Jpn Patent 06 256 405
77. Morita H, Hirota E, Ishizaki Y (1988) Euro Patent 273 605
78. Micale FJ (1987)US Patent 4 665 107
79. Yamaguchi S, Takabe T, Ito T, Kobayashi N, Morita H (1993) Jpn Patent 05 254 906
80. Yao H, Hayashi T (1992) Jpn Patent 04 300 946
81. Yao H, Hayashi T (1993) Jpn Patent 05287116
82. Giannelis EP, Krishnamoorti R, Manias E (1999) Adv Poly Sci 138:107 (and references therein)

83. McAvoy JM, Otaigbe JU (1997) Gas Atomization of polymers, SPE 55th ANTEC Papers (in press)
84. McAvoy JM (1997) MS Thesis Iowa State University
85. Jog JP (1993) Adv Polym Tech 12:281
86. Hoechst waxes for technical applications technical brochure (1993) Hoechst Celanese Corporation, Summit, New Jersey pp. 1–33
87. High-speed video tape available from the authors
88. Anderson IE, Figliola RS (1991) Adv Powder Metallurgy 5:237
89. Tuzun RE, Noid DW, Sumpter BG, Otaigbe JU (1997) Proc ACS:Div Polym Mat Sci 76:585
90. Chen CH, White JL, Spruiell JE, Goswami BC (1983) Textile Research Journal
91. Guandique E, Katz M (1964) US Patent 3,117,055
92. Kinney GA (1967) US Patent 3,338,992
93. Dees JR, Spruiell JE (1974) J Appl Polym Sci 18:1053
94. Sumpter BG, Getino C, Noid DW (1994) Annu Rev Phys Chem 45:439
95. Sumpter BG, Noid DW (1996) Annu Rev Mater Sci 26:223
96. Haykin S (1994) Neural networks:A comprehensive foundation. Macmillan New York
97. Hassoun MH (1995) Fundamentals of artificial neural networks. MIT Press Cambridge
98. Hagan MT, Demuth HB, Beale M (1996) Neural network design. PW Boston
99. Werbos PJ (1994) Roots of backpropagation:From ordered derivatives to neural networks and political forecasting Wiley New York
100. Rumelhart DE, McClelland JL (1986) Parallel distributed processing:Explorations in the microstructure of cognition. Vol 1 Foundations . MIT Press Cambridge
101. De Weijer AP, Buydens L, Kateman G, Heuvel HM (1992) Chemom Intell Lab Syst 16:77
102. Xi L (1991) EPD Congress pp 511–517
103. Smets HMG, Bogaerts WFL (1992) Mat Sel Des Sep:64
104. Joseph B, Hanratty FW, Kardos J (1995) J Composite Materials 29:100
105. May GS (1994) IEEE Spectrum 5:47
106. Bohr H (1988) IEEE Procedings pp 1034–1054
107. Chitra SP (1992) AI Expert 6:20
108. Chitra SP (1993) Chem Eng Progress 89:44
109. Hunt KJ, Sbarbaro, Zbikowski R, Gawthop PJ (1992) Automatica 28:1083
110. Fukuda T, Shibata T (1992) IEEE Trans Ind Electron.39:472
111. Thibault J, Grandjean BPA (1991) IFAC Symp Ser 8:251
112. Huang SH, Zhang HC(1994) IEEE Trans on Components, Packaging, and Manufacturing Technologies Part A, 17:212
113. Bulsari AB. (1994) J Systems Eng 4:131
114. Widrow B, Stearns SD (1985) Adaptive signal processing. Prentice-Hall Englewood Cliffs, NJ
115. Widrow B (1986) Proceedings of the Second IFAC Workshop on Adaptive Systems in Control and Signal Processing, Lund Institute of Technology Lund, Sweden, pp 1–5
116. Psaltis D, Sideris A, Yamamura AA (1988) IEEE Control Systems Magazine 8:17
117. Dirion JL, Cabassud M, Lann MVL, Casamatta G (1985) Computers Chem Engng 19:S797
118. Savkovic-Stevanovic J (1994) Computers Chem Engng 18:1149
119. Hornik K (1993) Neural Networks 6:1069
120. Hagan MT, Menhaj MB (1994), IEEE Trans Neural Networks 5:989
121. Polak E (1971) Computational methods in optimization. Academic New York
122. Powel MJD (1977) Math Program 12:241
123. Patrick P (1994) Neural Networks 7:1
124. Söderberg B (1988) Nucl Phys B 295, 396
125. HornikK, Stinchcombe M, White H (1989) Neural Networks 2:359
126. Masters T (1993) Practical neural network recipes in C++. Academic Press San Diego
127. Kohonen T (1995) Self-organizing maps. Springer-Verlag New York

128. Plutowski M, White H (1993) IEEE Trans Neural Networks 4:305
129. Efron B (1962) The jackknife, the bootstrap and other resampling plans Society for Industrial and Applied Mathematics Philadelphia
130. Stein R (1993) AI Expert February:42–47
131. Stein R (1993) AI Expert March:32–37
132. Crooks T (1992) AI Expert July:36–41
133. Perrone MP, Cooper LN (1992) Neural networks for speech and image processing. Chapman-Hall New York
134. Nguyen G, Widrow B (1989) Proc Int Joint Conf Neural Networks IEEE Press Washington D.C. pp. 21–26
135. Cybenko G (1989) Math Control, Signals Systems 2:303
136. Cybenko G (1988) Tech Rep Dept of Comp Sci, Tufts University, Medford,MA
137. Gori M, Tesi A (1992) IEEE Trans Pattern Anal Machine Intell PAMI-14:76
138. Broomhead K, Lowe D (1988) Complex Systems 2:321
139. Moody JE, Darken CJ (1989) Neural Comput 1:281
140. Renals S (1989) Electronics Letters 25:437
141. Poggio T, Girosi F (1990) Proceedings of the IEEE 78:1481
142. Leonard JA, Kramer MA, Ungar LH (1992).Computers Chem Engng 16:819
143. Jang JS, Sun C-T, Sun R (1993) IEEE Trans Neural Networks 4:156
144. Specht D (1991) IEEE Trans Neural Networks 2:586
145. Parzen E (1962) Annals Math Stat 33:1065
146. Masters T (1995) Advanced algorithms for neural networks:A C++ sourcebook. Wiley New York
147. Jenekhe SA, Zhang XJ, Chen XL Choong VE, Gao YL, Hsieh BR (1997) Chem Mater 9:409
148. Tarkka RM, Zhang XJ, Jenekhe SA (1996) J Am Chem Soc 118:9438
149. Croce F, Appetecchi GB, Persi L, Scrosati B (1998) Nature 394:456
150. Schwerzel RE, Spahr KB, Kurmer JP, Wood VE, Jenkins JA (1998) J Phys Chem A 102:5622
151. Xanthos M (1998) Poly Eng Sci 28:1392
152. Koning C, van Duin M, Pagnoulle C, Jerome R (1998) Prog Polym Sci. 23:707
153. Paul DR (1978) Polymer Blends, Paul DR, Newman S (ed) Academic Press,New York, pp 35–62
154. Utracki LA (1990) Polymer alloys and blends : Thermodynamics and rheology. Oxford University Press, New York
155. Sung L, Karim A, Douglas JF, Han CC (1996) Phys Rev Lett 76:4368
156. Yu JW, Douglas JF, Hobbie EK, Kim S, Han CC (1997) Phys Rev Lett. 78:2664
157. Marcus AH, Hussey DM, Diachun NA, Fayer MD (1996) J Chem Phys 103:8189
158. Jenekhe SA, Chen XL (1998) Science 279:1903 [see also Jenekhe SA, Chen XL (1999) ibid. 283:372]
159. Kung C-Y, Barnes MD, Lermer N, Whitten WB, Ramsey JM (1998) Analytical Chemistry 70:658
160. Kung C-Y, Barnes MD, Lermer N, Whitten WB, Ramsey JM (1999) Applied Optics 38:1481
161. We use the word "droplet" to denote liquid phase, and "particle" to denote a solid (dry) phase
162. Chang R, Davis EJ (1976) J Coll and Inter Sci 54:352
163. Davis EJ, Ray AK (1980) J Coll and Inter Sci 75:566
164. Ray AK, Souyri A, Davis EJ, Allen TM (1991) Appl Optics 30:3974
165. Widmann JF, Aardahl CL, Davis EJ (1996) Am Lab 28:35
166. See also, the review by E. J.Davis, Aer Sci Tech 26 212–254 (1997)
167. Chylek P, Ramaswamy V, Ashkin A, Dziedzic JM (1983) Appl Optics 22:2302
168. Van de Hulst HC, Wang RT (1991) Appl Optics 30:4755
169. Konig G, Anders K, Frohn A (1986) J Aerosol Sci 17:157

170. Glover AR, Skippon SM, Boyle RD (1995) Appl Optics, 34:8409
171. Widmann JF, Davis EJ (1996) Colloid and Polymer Science 274:525
172. Kaiser T, Lange S, Schweiger G (1994) Appl Optics 33:7789
173. Holler S, Pan Y, Chang RK, Bottinger JR, Hill SC, Hillis DB (1998) Optics Lett 23:1489
174. Widmann JF, Aardahl CL, Johnson TJ, Davis EJ (1998) J Coll Int Sci 199:197
175. Barnes MD, Lermer N, Whitten WB, Ramsey JM (1997) Rev Sci Instrum 68:2287
176. Lermer N, Barnes MD, Kung C-Y, Whitten WB, Ramsey JM (1997) Anal Chem 69:2115
177. Fukui K, Sumpter BG, Otaigbe JU, Barnes MD, Noid DW (1999) Macro Theory and Simulation 8:38
178. It should be noted that for heterogeneous particles, while the diffraction fringes are highly distorted, there remains some definite two-dimensional structure. This implies some uniformity and order of subdomains, however inversion of this type of data to extract such information is not trivial. This problem is currently under investigation. (See also Ref. 173).
179. Barnes MD, Ng KC, McNamara KP, Kung C-Y, Ramsey JM, Hill SC (accepted 1998) Cytometry (in press)
180. Lock JA (1990) Appl Optics 29:3180
181. Ray AK, Devakottai B, Souyri A, Huckaby JL (1991) Langmuir 7:525
182. Hightower RL, Richardson CB, Lin H-B, Eversole JD, Campillo AJ (1988) Optics Lett 13:946
183. Rigby D, Roe RJ (1987) J Chem Phys 87:7285
184. Rigby D, Roe RJ (1988) J Chem Phys 89:5280
185. Rigby D, Roe RJ (1989) Macromolecules 22:2259
186. Rigby D, Roe RJ (1990) Macromolecules 23:5312
187. Brown D, Clarke JHR (1991) Macromolecules 24:2075
188. Argon HAS, Suter UW (1993) Macromolecules 26:1097
189. Fan CF, Cagin T, Chen ZM, Smith KA (1994) Macromolecules 27:2383
190. Fan CF, Hsu SL (1992) Macromolecules 25:266
191. Sun Z, Morgan RJ, Lewis DN (1991) Polymer 33:2179
192. Noid DW, Pfeffer GA (1989) J. Polym. Sci. B 27:2321
193. Jin Y, Boyd RH (1998) J Chem Phys 108:9912
194. Yamamoto T (1997) J Chem Phys 107:2653
195. Sumpter BG, Noid DW, Wunderlich B (1990) J Chem Phys 93:6875
196. Neyertz S, Brown D, Thomas JO (1994) J Chem Phys 101:10064
197. Pant KPV, Boyd. RH (1993) Macromolecules 26:679
198. Han J, Boyd RH (1993) Macromolecules 27:5365
199. Fukuda M., Kuwajima S (1997) J Chem Phys 107:2149
200. Koike A (1998) Macromolecules 31:4605
201. Tillman PA, Rottach DR, McCoy JD, Plimton SJ, Curro JG (1997) J Chem Phys 107:4024
202. Grest GS, Lacasse M, Kremer K, Gupta AM (1996) J Chem Phys 105:10583
203. Hautman J, Klein M.L (1991) Phys Rev Lett 67:1763
204. Mar W, Klein M. (1994) J. Phys. Condensed Matter 6:A381
205. Mansfield KF, Theodorou D (1990) Macromolecules 23:4430
206. Mansfield KF, Theodorou D (1991) Macromolecules 24:6283
207. Yang X, Mao X (1997) Comput Theor Polym Sci 7:81
208. Liu C, Muthukumar M (1995) J Chem Phys 103:9053
209. Zhan Y, Mattice WL (1994) Macromolecules 27:7056
210. He D, Reneker DH, Mattice WL (1997) Comput Theor Polym Sci 7:19
211. Barnes MD, Lermer N, Kung CY, Whitten WB, Ramsey JM. (1997) Optics Lett 22:1265
212. Ichinose N, Ozaki Y, Kashu S (1992) Superfine Particle Technology, Springer-Verlag, London
213. Hayashi C, Uyeda R, Tasaki A (1997) Ultra-Fine Particles Technology, Noyes, New Jersey
214. Fukui K, Sumpter BG, Barnes MD, Noid DW, Otaigbe JU (1999) Macromol Theory Simul 8:28

215. Roy R, Sumpter BG, Pfeffer GA, Gray SK, Noid DW (1991) Physics Report 205:109
216. Roy R, Sumpter BG, Noid DW, Wunderlich B (1990) J Phys Chem 94:5729
217. Sumpter BG, Noid DW, Wunderlich B (1990) Polymer 31:1254
218. Sumpter BG, Noid DW, Wunderlich B, Cheng SZD (1990) Macromolecules 23:4671
219. Hoover WG (1983) Ann Rev Phys Chem 34:103
220. Klein ML (1985) Ann Rev Phys Chem 36:525
221. Noid DW, Sumpter BG, Wunderlich B, Pfeffer GA (1990) J Comp Chem 11:236
222. Gray SK, Noid, DW, Sumpter BG (1994) J Chem Phys 101:4062
223. Boyd RH, Breitling SM (1974) Macromolecules 7:855
224. Tanaka G, Mattice WL (1995) Macromolecules 28:1049
225. Michel A, Kreitmer S (1997) Comput Theor Polym Sci 7:113
226. Kavassalis TA, Sundarajan PR (1993) Macromolecules 26:4144
227. de Pablo JJ, Laso M, Suter UW (1992) J Chem Phys 96:2395
228. Martyna GJ, Klein ML, Tuckerman M (1992) J Chem Phys 97:2635
229. Fukui K, Cline JI, Frederick JH (1997) J Chem Phys 107:4551
230. Burkert U, Allinger NL (1982) Molecular Mechanics, American Chemical Society, Washington, D.C.
231. Connolly ML (1993) J Am Chem Soc 107:1118
232. Doucet J, Weber J (1996) Computer-Aided Molecular Design:Theory, Applications, Academic Press Inc., San Diego
233. Lewis M, Rees DC (1985) Science 230:1163
234. Wunderlich B (1990) Thermal Analysis, Academic Press, Inc., San Diego
235. Kreitmeier SN, Liang GL, Noid DW, Sumpter BG (1996) J Thermal Anal 46:853
236. Hor T, WadeAdams W, Pachter R, Haal P (1993) Polymer 34:2481
237. Crist B, Herena, PG (1996) J Polym Sci B Polymer Phys 34:449
238. Suhai S (1983) J Polym Sci Polym Phys Edu 21:1341
239. Meier RJ (1993) Macromolecules 26:4376
240. Hageman JCL, Meier RJ, Heinemann M, de Groot RA (1997) Macromolecules 30:5953
241. Shoemaker J, Horn T, Haal P, Pachter R, Adams WW (1992) Polymer 33:3351
242. Binder K (1999) Adv Poly Sci 138:1

Editor: Prof. J.E. McGrath
Received: April 2000

Dynamics in the Glassy State and Near the Glass Transition of Amorphous Polymers as Studied by Neutron Scattering

Toshiji Kanaya, Keisuke Kaji

Institute for Chemical Research, Kyoto University, Uji, Kyoto-fu, 611–0011, Japan
e-mail: kanaya@scl.kyoto-u.ac.jp

This review covers the recent progress of studies on dynamics in the glassy states and near the glass transition of amorphous polymers as revealed by inelastic and quasielastic neutron scattering, covering four topics: (i) low energy excitation (boson peak) in the glassy states; (ii) fast process in picosecond order; (iii) E-process related to conformational transitions; (iv) dynamical heterogeneity as revealed by non-Gaussian parameter.

Keywords. Amorphous polymers, Inelastic and quasielastic neutron scattering, Boson peak, Dynamics near T_g, Fast process, E-process, Heterogeneity

Advances in Polymer Science, Vol. 154
© Springer-Verlag Berlin Heidelberg 2001

List of Abbreviations and Symbols

A_0	non-Gaussian parameter
a_T	shift factor
b	scattering length
CRR	cooperatively rearranging region
D	diffusion coefficient
d	distance between two sites
E_a	activation energy
E_i, E_f	energy of incident and scattered neutron
e^{-2W_i}	Debye-Waller factor

f_ε	non-ergodic parameter
$G_{(\omega)}$	density of phonon (vibration) states
$G_s(r,t)$	time-space self-correlation function
$G_d(r,t)$	time-space distinct-correlation function
$G(r,t)$	time-space pair-correlation function
$g_G(<u^2>)$	Gaussian distribution function of mean square displacement
$h, \hbar$	Planck constant, $h/2\pi$
$I_{el}(Q)$	elastic neutron scattering intensity
$I_s(Q,t)$	incoherent intermediate scattering function
$I(Q,t)$	coherent intermediate scattering function
IXS	inelastic X-ray scattering
j_0	spherical Bessel function of zero order
$\mathbf{k}_i, \mathbf{k}_f$	wave vector of incident and scattered neutron
k_B	Boltzmann constant
KWW	Kohlrausch-WIlliams-Watts
$L(Q,\omega,\Gamma)$	Lorentzian function
MCT	mode coupling theory
MD	molecular dynamics
M	mass
m	fragility index
n_s	Bose-Einstein population factor
OTP	*ortho*-terphenyl
PB	*cis*-1,4-polybutadiene
PCP	*trans*-1,4-polychloroprene
PE	polyethylene
PET	poly(ethylene terephthalate)
PIB	polyisobutylene
PMMA	atactic poly(methyl methacrylate)
PS	atactic polystyrene
Q	scattering vector
Q	length of scattering vector
q	wave vector of phonon
r, ρ	position
$S(Q)$	structure factor
$S_{inc}(Q,\omega)$	incoherent scattering law
$S_{coh}(Q,\omega)$	coherent scattering law
$S_{vib}(Q,\omega)$	scattering law due to vibration
$S_{relax}(Q,\omega)$	scattering law due to relaxation
$S_{HN}(Q,\omega)$	scattering law from Havriliak-Negami function
T	absolute temperature
T_0	Vogel-Fulcher temperature
T_c	critical temperature in mode coupling theory
T_f	onset temperature of the fast process
T_K	Kauzmann Temperature
T_g	glass transition temperature

T_{gs}	glass transition temperature at slow cooling rate
T_{gf}	glass transition temperature at fast cooling rate
T_m	melting temperature
$U^j_\rho(q)$	polarization vector of the mode j, q
V_{SP}	specific volume
α, β, γ	exponent
ΔE	energy transfer
$\Delta I_{el}(Q)$	excess elastic scattering intensity
$\Delta I_{qes}(Q,\omega)$	excess quasielastic scattering intensity
$\Delta <u^2>$	excess mean square displacement
$\delta(\omega)$	Dirac delta function
ε	distance from critical temperature ($=T\text{-}T_c$)
$\Phi^*(\omega)$	Havriliak-Negami (HN) function
Γ	half-width at half-maximum of Lorentzian function
γ_r	friction term
Ω	solid angle
Θ	half of scattering angle
$\sigma_{inc}, \sigma_{coh}$	incoherent and coherent atomic scattering cross section
τ_o	average rest time of jump diffusion model
τ_1	average jump time of jump diffusion model
τ_f	characteristic time of the fast process
τ_r	mean residence time in two site model
τ_α	relaxation time of the α-process
τ_{JG}	relaxation time of the Johari-Goldstein process
ω	neutron frequency, also abbreviation of neutron energy $\hbar\omega$
ω_0	cutoff frequency
ω_b	boson peak frequency
ω_D	Debye frequency
$\omega_j(q)$	characteristic frequency of the mode j, q
$<\ell^2>$	mean jump distance of jump diffusion model
$<u^2>$	mean square displacement
$<u^2>_c$	mean square displacement of crystalline phase
$<u^2>_d$	mean square displacement of disordered phase

1
Introduction

Studies on polymer motions in solution and bulk have a long history, which revealed many modes of motion characteristic to polymers such as local conformational transitions, segmental motions, and reptation [1]. However, the original process causing these modes of motion are not well understood. In this review we try to visualize the origin of random motions in polymers, based on the experimental data from neutron scattering. On the other hand, dynamics of glass-forming polymers has been extensively studied in the last decade from a viewpoint of dynamics of glass-forming materials including inorganic and organic molecules as well as polymers, using many kinds of experimental techniques [2–6]. These studies revealed many anomalous but common features of dynamics of glass-forming materials in glassy states and in supercooled states. These experimental studies have been stimulated by some theories, especially by the mode coupling theory [7], which shed light on the substantial nature of the glass transition phenomenon. It was also found through these studies that some of the dynamic features of amorphous polymers must be considered to be common features of glass-forming materials, and alternatively, some of them to be a special feature of polymer motions.

Many of the experimental studies have been carried out using microscopic methods such as inelastic and quasielastic neutron scattering, inelastic X-ray scattering, dynamic light scattering, wide band dielectric relaxation measurements, and NMR although macroscopic measurements such as mechanical relaxation and thermal measurements were as well used for investigations of glass transition, especially in the earlier stages. Among microscopic measurements, the contribution of neutron scattering techniques is outstanding because results from these areas can be compared directly with theoretical predictions and the results of computer simulations.

In this article we review recent investigations into dynamics of amorphous polymers below and above the glass transition temperature T_g using inelastic and quasielastic neutron scattering techniques. As mentioned above, these studies have revealed some common dynamic features not only of glass-forming polymers but also of other types of glass-forming materials. To illustrate general features, we will cite many experimental results on organic and inorganic glass-forming molecules throughout the article.

The arrangement of this article is as follows. In Sect. 2, we briefly describe basic principles of inelastic and quasielastic neutron scattering for the readers' convenience. In Sect. 3, after showing anomalous thermal properties of glass-forming materials at low temperatures, we review recent studies on an anomalous low-energy excitation, the so-called boson peak, observed for most glass-forming materials. Section 4 is dedicated to dynamics near the glass transition, especially focusing on the fast process in picosecond order. Local conformational transitions in a polymer chain is discussed in Sect. 5 on the basis of the E-process observed by quasielastic neutron scattering in a time range of several tens to several hundreds of picoseconds, a subject characteristic of polymer dynamics. In Sect. 6, dynamic heterogeneity in amorphous polymers is discussed in terms of non-Gaussian parameters evaluated from the Q dependence of incoherent elastic scattering intensity. Finally, concluding remarks are given in Sect. 7.

2
Inelastic and Quasielastic Neutron Scattering

There are many textbooks dealing with principles of neutron scattering and its applications to various fields [8–13], some of which [14] focus on polymer studies. Hence, in this section we outline only some essential basic principles of inelastic and quasielastic neutron scattering to help readers understand the contents of this review.

2.1
Momentum Transfer and Energy Transfer

In Fig. 1 the neutron scattering process in real and momentum space is schematically illustrated. In experiments, we observe double-differential scattering cross section $\partial^2\sigma/\partial\Omega\partial E$, which is a probability that an incident neutron with wave vector k_i and energy E_i is scattered by a scattering angle 2Θ into a solid angle element $d\Omega$ and an energy interval between E_f and E_f+dE. The energy and wave vector of the scattered neutron are E_f and k_f, respectively. The momentum transfer $\hbar Q$ and energy transfer ΔE of the neutron are given by

$$\hbar Q = \hbar(k_f - k_i) \tag{1}$$

$$\Delta E = \hbar\omega = E_f - E_i = \frac{\hbar}{2m}(k_f^2 - k_i^2). \tag{2}$$

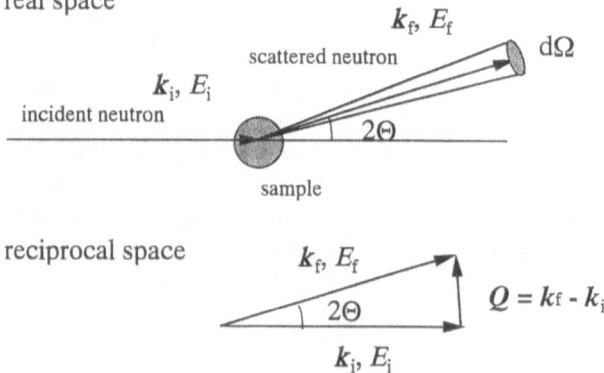

Fig. 1. Neutron scattering process in real and momentum space. An incident neutron with wavenumber k_i and energy E_i is scattered by a scattering angle 2Θ into a solid angle element $d\Omega$. Wave number and energy of the scattered neutron are k_f and E_f, respectively

2.2
Double-Differential Scattering Cross-Section, Dynamic Scattering Law, and Intermediate Scattering Function

For a system consisting of one kind of atom the double-differential scattering cross-section $\partial^2\sigma/\partial\Omega\partial E$ is simply related to the scattering law

$$\frac{\partial^2\sigma}{\partial\Omega\partial E} = \frac{k_f}{4\pi\hbar k_i} N\left[\sigma_{inc}S_{inc}(Q,\omega) + \sigma_{coh}S_{coh}(Q,\omega)\right] \tag{3}$$

where $S_{inc}(Q,\omega)$ and $S_{coh}(Q,\omega)$ are incoherent and coherent scattering laws, σ_{inc} and σ_{coh} are incoherent and coherent atomic scattering cross sections, respectively, and N is number of atoms. The incoherent and coherent dynamic scattering laws are related to incoherent and coherent intermediate scattering functions $I_s(Q,t)$ and $I(Q,t)$, respectively, through the Fourier transform with ω:

$$S_{inc}(Q,\omega) = \frac{1}{2\pi}\int_{-\infty}^{+\infty} I_s(Q,t)\exp(-i\omega t)dt \tag{4}$$

$$S_{coh}(Q,\omega) = \frac{1}{2\pi}\int_{-\infty}^{+\infty} I(Q,t)\exp(-i\omega t)dt \tag{5}$$

Hence, the incoherent and coherent intermediate scattering functions $I_s(Q,t)$ and $I(Q,t)$ are respectively defined by

$$I_s(Q,t) = \frac{1}{N}\sum_k \left\langle \exp\{-iQr_k(0)\}\exp\{iQr_k(t)\} \right\rangle \tag{6}$$

$$I(Q,t) = \frac{1}{N}\sum_k\sum_l \left\langle \exp\{-iQr_k(0)\}\exp\{iQr_l(t)\} \right\rangle. \tag{7}$$

2.3
Time-Space Correlation Function

Following the van Hove treatment [15], incoherent and coherent dynamic scattering laws $S_{inc}(Q,\omega)$ and $S_{coh}(Q,\omega)$ are given by the time-space Fourier transforms of the time-space self- and pair-correlation functions $G_s(r,t)$ and $G(r,t)$, respectively:

$$S_{inc}(Q,\omega) = \frac{1}{2\pi}\int_{-\infty}^{+\infty}\int_0^{+\infty} drdt G_s(r,t)\exp[i(Qr-\omega t)] \tag{8}$$

$$S_{coh}(Q,\omega) = \frac{1}{2\pi}\int_{-\infty}^{+\infty}\int_0^{+\infty} drdt G(r,t)\exp[i(Qr-\omega t)]. \tag{9}$$

It is useful to separate the pair-correlation function $G(r,t)$ into a self and distinct part, namely

$$G(r,t) = G_s(r,t) + G_d(r,t) \tag{10}$$

Assuming that there is a particle at a position $r=0$ at time $t=0$, the physical meaning of the self-correlation functions $G_s(r,t)$ is a probability of finding the same particle in dr at a position r at a later time t. In analogy, the distinct-correlation function is a probability of finding another particle at r at a time t.

2.4
Gaussian Approximation

If the distribution of particle displacement [$r_k(t)-r_k(0)$ or $r_k(t)-r_l(0)$] is described by a Gaussian distribution (the so-called Gaussian approximation), incoherent and coherent intermediate scattering functions can respectively be written in terms of only Q^2 as

$$I_s(Q,t) = \frac{1}{N}\sum_k \exp\{-\frac{Q^2}{6}\langle(r_k(t)-r_k(0))^2\rangle\} \tag{11}$$

$$I(Q,t) = \frac{1}{N}\sum_k\sum_l \exp\{-\frac{Q^2}{6}\langle(r_k(t)-r_l(0))^2\rangle\} \tag{12}$$

Here it is worth pointing out that the time dependent mean square displacement is observed from the incoherent scattering measurements. It is also noted that the frequency dependent mean square displacement can be obtained from the measurements of incoherent dynamic scattering law (Eq. 4) within the Gaussian approximation.

2.5
Coherent and Incoherent Cross-Sections

In most cases, samples are a mixture of isotopes j with different scattering length b_j and it is assumed that they are randomly distributed over the sample. Such random distribution of different isotopes produces incoherent cross-section. In analogy with isotope incoherence, spin incoherence is also observed. For nuclei with spin I≠1, the interaction depends on the orientation between neutron and nuclear spins; scattering lengths b_+ and b_- for parallel and antiparallel spin, respectively, are different and the orientations of spins are randomly distributed in the nuclei even if they are the same kinds of nuclei. Then, incoherent and coherent atomic cross-sections are given by

$$\sigma_{inc} = 4\pi(<b^2> - <b>^2) \tag{13}$$

$$\sigma_{coh} = 4\pi <b>^2 \tag{14}$$

where $<b^2>$ and $<b>^2$ are the mean square length and the square of mean scattering length, respectively. For organic polymers which usually include many hydrogen atoms, contribution from spin incoherence is very large (for H atom, σ_{inc} and σ_{coh} are 79.9 and 1.8 barns [16], respectively). Therefore, incoherent scattering is dominant for many organic polymers at large momentum transfers.

2.6
Inelastic Scattering from Phonons

Inelastic scattering of neutrons is caused by an oscillatory motion. An example is the inelastic scattering of neutrons by phonons. Following a textbook [10], the coherent double-differential scattering cross-section of neutrons for one phonon process is given by

$$\frac{\partial^2 \sigma_{coh}^1}{\partial \Omega \partial E} = \frac{(2\pi)^3}{v} \sum_{q,j} \frac{k_f}{k_i} \delta(\hbar\omega \mp \hbar\omega_j(q)) \sum_{\tau} \delta(Q \mp q - \tau)$$

$$x \frac{\hbar(n_s + \frac{1}{2} \pm \frac{1}{2})}{2\omega_j(q)} \left| \sum_{\rho} \frac{<b>_\rho}{M_\rho} e^{iQr} Q \cdot U_\rho^j(q) e^{-2W_\rho} \right|^2 \tag{15}$$

Here, q is the wave vector of phonon, $\omega_j(q)$ is the characteristic frequency of the mode j at q and τ is the reciprocal lattice vector. $<b>_\rho$ and M_ρ are the coherent scattering length and the mass of the atom at ρ in the unit cell, respectively, and e^{-2W_ρ} is its Debye-Waller factor. $U_\rho^j(q)$ is the polarization vector, corresponding to the displacement vector of a simple harmonic oscillator. The upper and the lower of double signs in the equation refer to neutron energy loss or phonon creation and to neutron energy gain or phonon annihilation, respectively. All phonon modes are harmonic oscillators, and hence they have Bose-Einstein population factors $n_s + 1$ for energy loss and n_s for energy gain:

$$n_s = [\exp(\hbar\omega_j(q)/k_B T) - 1]^{-1} \tag{16}$$

Two δ-functions in Eq. (15) mean that scattering occurs under the conditions of energy and momentum conservation. One obtains a phonon peak in experiments, not phonon distributions.

The incoherent double-differential scattering cross-section for the one phonon process is given by

$$\frac{\partial^2 \sigma_{inc}^l}{\partial \Omega \partial E} = \sum_{q,j} \frac{k_f}{k_i} \delta(\hbar\omega \mp \hbar\omega_j(q)) \frac{\hbar(n_s + \frac{1}{2} \pm \frac{1}{2})}{2\omega_j(q)}$$

$$x \sum_{\rho} \frac{(<b^2>_\rho - <b>^2_\rho)}{M_\rho} \left| Q \cdot U_\rho^j(q) \right|^2 e^{-2W_\rho} \tag{17}$$

$4\pi(<b^2>_\rho - <b>^2_\rho)$ corresponds to incoherent atomic scattering cross section. Equation (17) includes one δ-function for energy conservation, meaning that the scattering consists of a broad distribution in energy. For a special case of a crystal with cubic symmetry, containing one atom in the unit cell, the relation

$\left\langle (Q \cdot U)^2 \right\rangle = (1/3)Q^2 <u^2>$ is sustained where $<u^2>$ is the mean square amplitude of vibration of the atom. Assuming continuous distribution of the frequency, Eq. (17) is reduced to

$$\frac{\partial^2 \sigma_{inc}^l}{\partial \Omega \partial E} = \frac{\hbar k_f}{k_i} (<b^2> - <b>^2) \frac{Q^2 <u^2>}{2M} (n_s + \frac{1}{2} \pm \frac{1}{2}) e^{-2W} N \frac{G(\omega)}{\omega} \tag{18}$$

where $G(\omega)$ is the normalized density of phonon states and $3N$ is the total number of phonon states.

2.7
Quasielastic Scattering from Some Models

Quasielastic scattering is caused by random motions or energy dissipation processes. One of the typical examples of such motions is a simple diffusion described by the Fick's law. The incoherent intermediate scattering function is given by

$$I_{inc}(Q,t) = \int_0^\infty G_s(r,t)\exp(-Q \cdot r)dr = \exp(-DQ^2 t) \tag{19}$$

where D is a diffusion coefficient. The corresponding incoherent scattering law is given by a Lorentzian function whose half-width at half-maximum increases with the momentum transfer according to a DQ^2 law:

$$S_{inc}(Q,\omega) = \frac{1}{\pi} \frac{DQ^2}{\omega^2 + (DQ^2)^2} \tag{20}$$

In cases of real systems random motions are often limited in a finite space. Diffusion motions in a sphere and jump motions between two sites [13] are such cases. For the jump motion between two equivalent sites separated by a distance d with the mean residence time τ_r in each site, the scattering law after a power average is given by

$$S(Q,\omega) = F_0(Q)\delta(\omega) + F_1(Q)\frac{1}{\pi}\frac{2\tau_r}{4+\omega^2\tau_r^2} \qquad (21)$$

$$
\begin{aligned}
F_0(Q) &= [1 + j_0\{Qd\}]/2 \\
F_1(Q) &= [1 - j_0\{Qd\}]/2
\end{aligned}
\qquad (22)
$$

where j_0 is a spherical Bessel function of zeroth order. Due to the spatial restriction, the dynamic scattering law $S(Q,\omega)$ includes a δ-function, giving a geometrical information of the motion.

The segmental motion of a polymer chain was successfully described by a bead-spring model, discussed by Rouse [17] in the so-called free-draining limit and by Zimm [18] in the hydrodynamic limit. de Gennes [19, 20] calculated the coherent and incoherent intermediate scattering functions for both the Rouse and Zimm models. In the low Q and long time limit, the time decay of the intermediate scattering function depends on $t^{1/2}$ and $t^{2/3}$ and the Q dependence of the decay rate or the half-width of the corresponding scattering law follows Q^4 and Q^3 law for the Rouse and Zimm model, respectively.

3
Low-Energy Excitation in Amorphous Polymers (Boson Peak)

3.1
Anomalies of Thermal Properties

Anomalous thermal properties of amorphous materials at low temperatures [21] are one of the unsolved problems in solid state physics as well as in polymer physics, though much effort has been made by many researchers [2–6]. A typical manifestation of the anomalies is heat capacity. As pointed out by Zeller and Pohl [22], heat capacities of amorphous materials show an excess value compared with those of crystalline materials in two low temperature regions; one is below 1 K where heat capacity is proportional to temperature T, the other is an excess heat capacity at around 10~20 K. This is schematically illustrated in Fig. 2. The former has been interpreted in terms of a tunneling motion in an asymmetric double well potential proposed by Phillips [23] and Anderson et al. [24] independently, but the latter is less understood. Recently, extensive studies on the latter have been performed by inelastic neutron scattering [25–39] and Raman scattering techniques [40–47]. These studies have revealed that an excess excitation peak exists at 2~3 meV in spectra of all amorphous materials well below the glass transition temperature T_g, corresponding to the excess heat capacity at around 10~20 K. This peak is called the "boson peak." In this section we review the recent neutron scattering investigations about the boson peak, especially focusing on amorphous polymers.

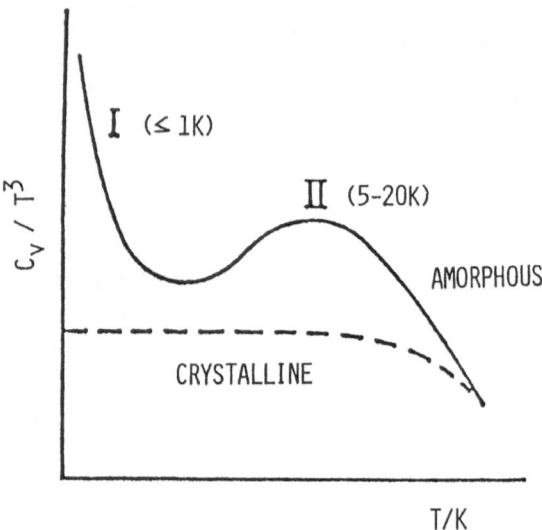

Fig. 2. Schematic representation of temperature dependence of heat capacity divided by T^3 (C/T^3) for amorphous and crystalline materials at low temperature regions

3.2
Universality of Low-Energy Excitations

Thermal anomalies at low temperatures around 10~20 K are observed for all amorphous materials, suggesting that the boson peak should be observed in all amorphous materials. In order to confirm this, inelastic neutron scattering measurements were carried out on various amorphous polymers as well as inorganic glasses [30, 34]. Observed dynamic scattering laws $S(Q,\omega)$ are shown for ten organic polymers and three inorganic glasses in Fig. 3. Organic polymers used for the scattering experiments are polyisobutylene (PIB), *cis*-1,4-polybutadiene (PB), cross-linked *cis*-1,4-polybutadiene (cl-PB), *trans*-1,4-polychloroprene (PCP), atactic polystyrene (PS), highly crystalline polyethylene (h-PE) with a degree of crystallinity 0.96, semicrystalline polyethylene (s-PE) with a degree of crystallinity 0.46, poly(ethylene terephthalate) (PET), poly(methyl methacrylate) (PMMA), and epoxy resin (EXPO) cured with polyamide amine. Three inorganic glasses of germanium (GeO_2) glass, boric oxide (B_2O_3) glass, and amorphous selenium (Se) are also used. All the samples except highly crystalline polyethylene (h-PE) show a broad excitation peak in the energy range of 1.5~4.0 meV. These data strongly demonstrate that the boson peak is a universal property for amorphous materials.

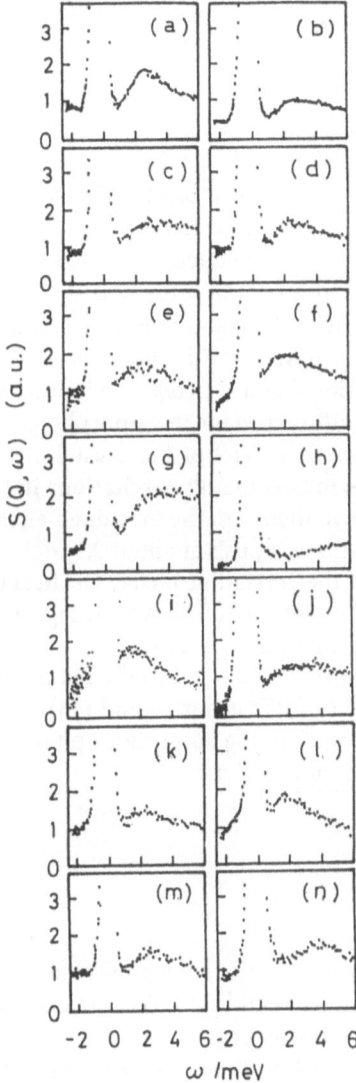

Fig. 3a–n. Dynamic scattering laws $S(Q,\omega)$ for various amorphous polymers: **a** polyisobutylene (PIB) at 50 K; **b** PIB at 10 K; **c** *cis*-1,4-polybutadiene (PB) at 50 K; **d** cross-linked PB at 50 K; **e** *trans*-1,4-polychloroprene (PCP) at 50 K; **f** atactic polystyrene (PS) at 10 K; **g** semicrystalline polyethylene (s-PE) with degree of crystallinity of 0.46 at 10 K; **h** highly crystalline polyethylene (h-PE) with degree of crystallinity of 0.96 at 10 K; **i** poly(ethylene terephthalate) (PET) at 18 K; **j** atactic poly(methyl methacrylate) (PMMA) at 18 K; **k** epoxy resin (EXPO) at 50 K; **l** amorphous selenium (a-Se) at 150 K; **m** germanium (GeO_2) glass at 295 K; **n** boric oxide (B_2O_3) glass at 295 K. Spectra of (a)–(k) are obtained by summing up the seven spectra at scattering angles at 8°, 24°, 40°, 56°, 72°, 88°, and 104°. Spectra of (l)–(n) are at $Q=2.07$ Å^{-1}. (Reprinted with permission from [34]. Copyright 1994 Elsevier Science B. V., Amsterdam)

3.3
Crystalline and Amorphous Phases

In order to show the differences between the low energy excitations of amorphous and crystalline phases, we focus on the results of polyethylenes (PE) with different degrees of crystallinity [29, 34] (see Fig. 3g,h). Assuming that the observed $S(Q,\omega)$ is described by a linear combination of those of the crystalline and amorphous phases, the dynamic scattering laws $S(Q,\omega)$ of the crystalline and amorphous phases of PE were evaluated and shown in Fig. 4a. In the spectrum of the amorphous phase, a broad peak is observed at about 3 meV. This broad peak is absent in the spectrum of the crystalline phase. It is then directly concluded that the boson peak at around 3 meV is characteristic of the amorphous phase.

The density of vibrational states $G(\omega)$, which was calculated from $S(Q,\omega)$, is plotted in the form $G(\omega)/\omega^2$ vs ω for the amorphous and crystalline phases in Fig. 4b. The figure shows that $G(\omega)$ of the crystalline phases is *approximately* proportional to ω^2. This means that the excitations in the crystalline phase can be described by the Debye theory in the examined energy range. For the amorphous phase, $G(\omega)/\omega^2$ shows a peak at about 2.5 meV and the value of $G(\omega)$ is much larger than that of the crystalline phase. The heat capacity $C(T)$ was calculated from the density of states $G(\omega)$ shown in Fig. 4b for the amorphous and crystalline phases in a temperature range of 2–15 K and the results are plotted as $C(T)/T^3$ in Fig. 4c. The agreement between the temperature dependences of the calculated and observed $C(T)/T^3$ is fairly good [48]. It is confirmed that the boson peak at around 2~3 meV of the amorphous phase is the origin of the excess heat capacity of the amorphous phase.

3.4
Properties of Boson Peak

In this section we briefly summarize some typical features of the boson peak observed in amorphous materials and some recent experimental results which give hints for the origin of the boson peak.

As shown in Sect. 3.2, the boson peak appears irrespective of different chemical structures of amorphous materials. However, the details in energy, shape and intensity of the peak depend on the chemical structures and temperature. Studies on partially deuterated polystyrene [31] revealed that the boson peak of the main chain (methylene chain) appears at 1.5 meV, which is 0.7 meV lower than the value of 2.2 meV for the side chain (phenyl group). This lower energy shift was assigned to the larger effective mass of the main chain because the main chain motion may be the motion of several whole monomeric units. In Fig. 5, the monomer mass dependence of the boson peak energy ω_b is shown for various amorphous polymers. The boson peak energy ω_b decreases with increasing mass of monomer while it becomes almost independent of mass above a certain value (see Fig. 5). Similarly, the boson peak energy ω_b for van der Waals

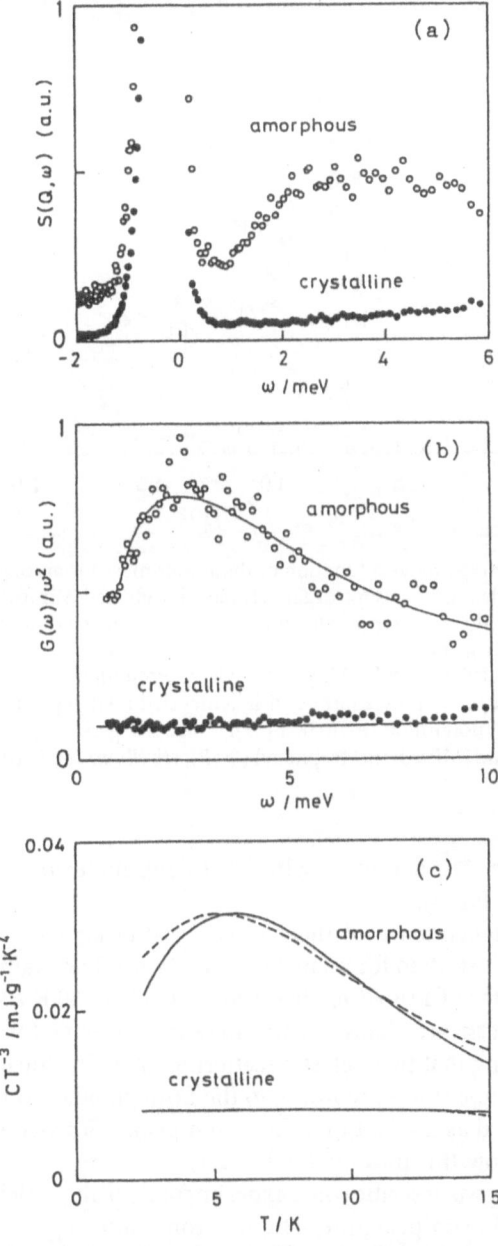

Fig. 4.a Dynamic scattering laws $S(Q,\omega)$ for amorphous and crystalline phases of PE at 10 K. **b** Density of states $G(\omega)$ for amorphous and crystalline phases of PE plotted as $G(\omega)/\omega^2$ against ω. *Solid lines* are fitting curves. **c** Heat capacity $C(T)$ for amorphous and crystalline phases of PE. *Solid lines* were calculated from the fitting curves of $G(\omega)$ in (b) and *dashed lines* are from experimental data [48]. (Reprinted with permission from [34]. Copyright 1994 Elsevier Science B. V., Amsterdam)

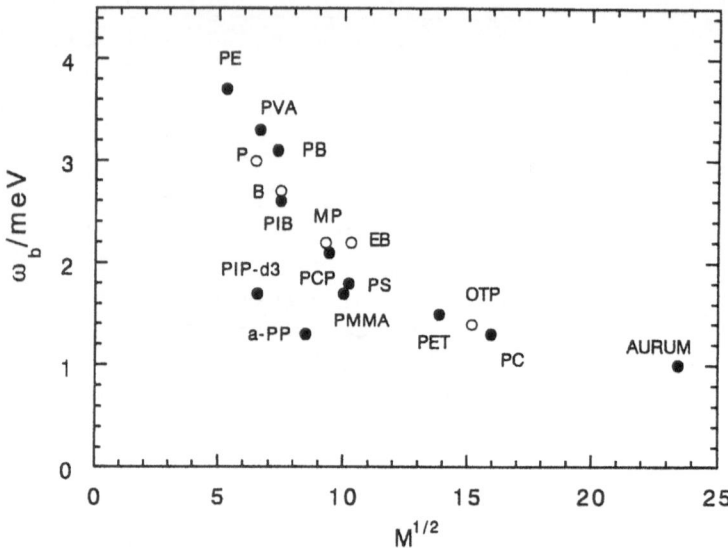

Fig. 5. Boson peak energy ω_b as a function of monomer mass for amorphous polymers and as a function of molecular mass for organic molecular glasses. Amorphous polymers: polyethylene (PE) [29], poly(vinyl alcohol) (PVA) [152], *cis*-1,4-polybutadiene (PB), polyisobutylene (PIB), *trans*-1,4-polychloroprene (PCP), atactic polystyrene (PS), atactic poly(methyl methacrylate) (PMMA), poly(ethylene terephthalate) (PET) [30, 34], atactic polypropylene (a-PP) [95], polyisoprene with deuterated methyl group (PIP-d$_3$) [36], polycarbonate (PC) [91], polyimide (AURUM) [39]. Molecular glasses: propylene (P), 1-butene (B), 3-methyl pentane (MP), ortho-terphenyl (OTP), ethylbenzene (EB) [38]

molecular glasses [38] decreases with increasing molecular weight below the certain value (see Fig. 5).

Temperature dependence of the inelastic scattering intensity of the boson peak has been well studied for many kinds of glasses. For example, the densities of vibrational states of PIB $G(\omega)$ observed at 10 K and 50 K are shown in Fig. 6 [30]. Both the spectra are identical when they are corrected for the Bose population factor, showing that the inelastic scattering intensity due to the boson peak increases with temperature according to the Bose population factor. This result is often interpreted as a result that the boson peak is a harmonic vibration. As will be shown below, this may not be the case.

In the following we describe some experimental results which give some hints for the origin of boson peak. Inelastic neutron scattering measurements were performed on amorphous and semicrystalline poly(ethylene terephthalate)s (PET) to evaluate dynamic scattering laws $S(Q,\omega)$ of the pure amorphous and crystalline contributions [49]. In contrast to the results of PE [29], the $S(Q,\omega)$ of the crystalline phase of PET shows a clear boson peak while the intensity is slightly lower than that of the amorphous phase, suggesting that dynamic defects in the crystalline phase are a possible origin of the boson peak.

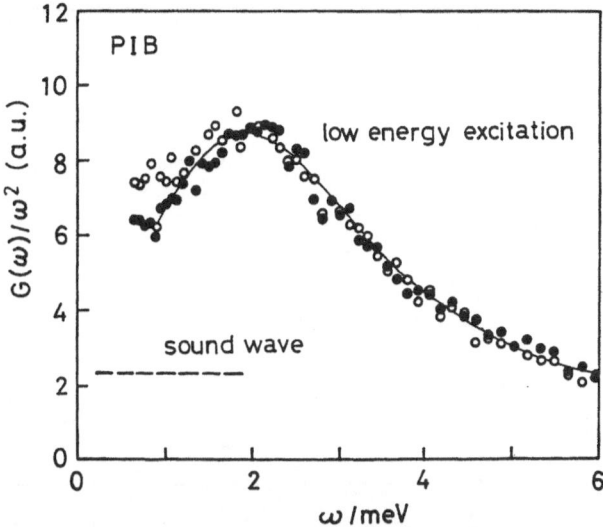

Fig. 6. Density of vibrational states $G(\omega)$ of polyisobutylene (PIB) measured at 10 K and 50 K. The two $G(\omega)$s are identical, suggesting that the temperature dependence of inelastic scattering intensity is governed by the Bose population factor. (Reprinted with permission from [30]. Copyright 1991 American Institute of Physics, New York)

Structurally disordered and orientationally disordered glassy states of ethanol were studied by incoherent neutron scattering by Ramos et al. [50]. They showed that the orientationally disordered glassy state still keeping translational order already shows the boson peak with an intensity of more than about 90% of that of the structurally disordered glassy states. This result indicates that the orientational disorder also gives a boson peak. Further inelastic neutron measurements were performed on macroscopically oriented amorphous PET fiber in two scattering configurations: Q parallel and perpendicular to the orientational axis, resulting in both the observed dynamic scattering laws $S(Q,\omega)$ being almost identical [51]. This data suggests that the orientational parts have no effect on the boson peak. The reason for this is not clear at the moment.

From another point of view, annealing effects on the boson peak of poly(methyl methacrylate) (PMMA) [52] were studied using inelastic neutron scattering to find that the boson peak intensity decreases by annealing at a temperature slightly below the glass transition temperature as shown in Fig. 7. The intensity decreases in the energy range below about 3 meV while in a high energy region it does not change, suggesting that the structure related to the boson peak can relax with annealing.

One of the challenging works is to evaluate the spatial scale of boson peak mode. By analyzing the Q dependence of coherent inelastic neutron scattering from vitreous silica, Buchenau et al. [26] assigned the boson peak mode to a coupled rotation of five tetrahedra of SiO_4, which correspond to a spatial scale of

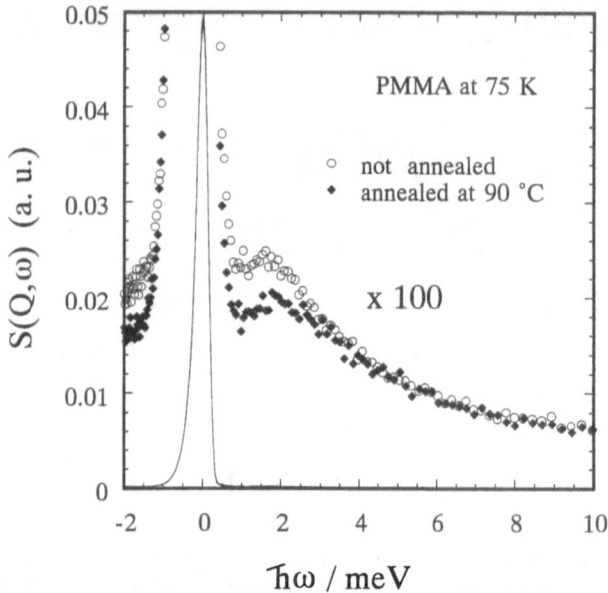

Fig. 7. Dynamic scattering laws $S(Q,\omega)$ of poly(methyl methacrylate) (PMMA), not annealed (○) and annealed at 90 °C (◆). The measurements were done at 75 K

10~20 Å. Incoherent inelastic neutron scattering measurements [53] gave some evidences that the boson peak is a localized mode on 6 and 11 monomers for PS and PB, respectively. Such localization was predicted in molecular dynamics simulations [54–56]. In the framework of the soft potential model, which will be mentioned later, the number of atoms participating in the boson peak mode was evaluated, showing the localization of the boson peak mode on several tens to hundreds of atoms for three inorganic glasses [28] and on several monomers for three polymers [57].

Finally, we have to mention recent inelastic X-ray scattering (IXS) measurements on some glasses [58–60]. X-ray scattering gives coherent scattering and is accessible to a lower Q range compared with neutron scattering. Due to such characteristic features, IXS revealed an extended mode in the boson peak energy region. In the beginning there was a discussion that the boson peak is an extended mode, but now it is clear that this is wrong [61, 62]. At present it is considered that an extended mode (Debye mode) and a localized mode (boson peak) coexist in the low energy region.

3.5
Some Models

To explain the origin of the boson peak, some models or theories were proposed, and we will examine some of them in this section.

A fashionable explanation for the boson peak is by "fracton." The fracton theory [63] predicted that excess excitation exists in a crossover energy range between phonon and fracton. This idea was successfully applied to the density of states of cross-linked epoxy resin [64, 65]. However, computer simulation of large fractal networks by Yakubo et al. [66] showed that there are no excess excitations in the crossover region. It is often assumed that the origin of the boson peak is a confined phonon [46, 47, 67]. For amorphous polymers it was assumed that a glassy state has a non-continuous structure due to a stronger bonding between atoms in blobs than between atoms in different nearest-neighbor blobs. Such localized vibration mode in the blob has a lowest frequency ω_0 and is related to the lowest energy of the boson peak.

The most plausible explanation at present is based on an idea of strain induced mode softening. Generally speaking, the potential in crystalline state is described to be harmonic while that in glassy state must be deformed due to disordered strain, leading to softening of the potential. It sometimes gives a negative force constant. This idea was reasonably manifested in the soft potential model by Buchenau et al. [28, 68, 69]. Within the soft potential model they calculated density of tunneling and vibrational states using a specific assumption on the asymmetry of the soft potential and found that the model can well reproduce the temperature dependence of heat capacity as well as thermal conductivity for amorphous silica and amorphous selenium.

As mentioned in Sect. 3. 4, recent IXS experiments have revealed that there coexist the boson peak and an extended mode in the low energy region. This was observed in MD simulation of a vitreous silica model [61] and theoretically explained by a model proposed by Nakayama [62, 70].

4
Fast Processes in Amorphous Polymers near T_g

In addition to the low energy excitations in glassy polymers, the dynamics near the glass transition is also one of the current topics in this field [2–6, 71]. Glass transition is easily detected using many experimental techniques such as volumetry, thermal analysis, mechanical or dielectric dispersion, and spectroscopic methods. A schematic representation of specific volume measurements of a glass-forming material is given in Fig. 8. When a glass-forming material is cooled down from a high temperature above the melting temperature T_m, it easily gets into a supercooled state without any drastic changes at T_m. With further decreasing temperature, the thermal expansion coefficient changes at a certain temperature. This is a glass transition temperature T_g. It is known that T_g depends on the cooling rate, and hence the glass transition is considered a relaxation process or dynamical process. Therefore, extensive studies have been performed on the dynamics of glass-forming materials.

A microscopic theory, the so-called mode coupling theory (MCT) [7, 72], has been developed recently, based on an equation for the density autocorrelation function which contains a nonlinear memory function and gives some detailed

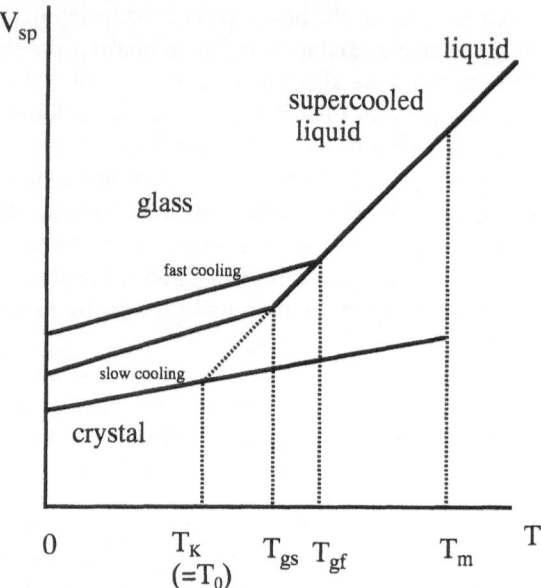

Fig. 8. Schematic representation of temperature dependence of specific volume of a glass-forming material. Thermal expansion coefficient changes at the glass transition temperature T_g which depends on cooling rate. The faster the cooling rate the higher the glass transition temperature T_g

pictures and predictions for the dynamic process involved in the glass transition. Stimulated by this theory, many experimental works have been performed on inorganic [73, 74] and organic [75–78] glass-forming materials, including polymers [79–83], using neutron scattering, light scattering, and dielectric relaxation techniques. These works revealed anomalous but common features of the dynamics near T_g. One of the most outstanding results is the finding of a very fast motion of picosecond order appearing below T_g in various glass-forming materials in an energy range below about 2 meV [75–77]. It is obvious that the so-called α-process is mainly concerned with the glass transition phenomena. However, it is also essential to elucidate the fast process of picosecond order near T_g in order to understand the glass transition dynamics.

In this section we review the recent results of the studies on glass transition dynamics revealed by microscopic experimental methods, especially by neutron scattering techniques, and discuss roles of the newly found fast process in the glass transition dynamics.

4.1
Overview of Dynamics Near T_g

One of the fruitful results of the extensive studies on dynamics of glass-forming materials near T_g in the last decade is relaxation time maps of some glass-forming polymers. We give an overview of the glass transition dynamics on the basis of the relaxation time map of amorphous polybutadiene (PB).

4.1.1
Relaxation Time Map

The characteristic times τ of PB determined by many experimental methods are plotted in a relaxation time map (Fig. 9), where $1/\tau$ is plotted vs inverse temperature $1/T$. This map was first reported by Rössler et al. [84] to summarize the characteristic features of the dynamics of glass-forming materials.

At a frequency around 10^{12} Hz, the fast process is observed, whose relaxation time does not depend on temperature [32,79]. Such a fast process was first found for some polymers by Kaji et al. [85–87]. Since then, similar fast processes have been observed in inorganic and organic glass-forming molecules [75–78] as well as in many glass-forming polymers [32, 35, 36, 39, 51, 79, 88–95]. The dynamic properties are rather similar to those of the β-process predicted by the MCT, so that it was often discussed on the basis of MCT [77, 96–98]. However, the origin is not yet clarified. The present situation of understanding of the fast process, which is one of the main subjects of this review, will be described later.

Below the fast process in frequency, the so-called primary α-process is observed [80, 99] which governs the viscosity [100] and is directly related to the glass transition. The temperature dependence of the relaxation time τ is described not by the Arrhenius equation but by the Vogel-Fulcher type:

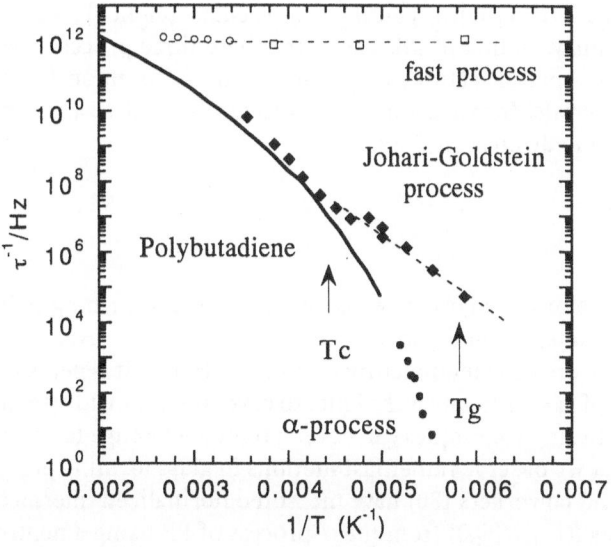

Fig. 9. Relaxation time map of polybutadiene observed with various kinds of methods. Quasielastic neutron scattering (O, □) [32, 79], neutron spin echo (◆) [80, 82], viscosity (●) [100], ^{2}H NMR (l) [99]

$$\log \tau = \log \tau_\infty + \frac{B}{T - T_0} \qquad (23)$$

where T_0 ($\approx T_g$–50) is called Vogel-Fulcher temperature, and τ_∞ and B are constants. This is identical to the Williams-Landel-Ferry (WLF) equation [101]. The meaning of Eq. (23) is that an apparent activation energy increases with decreasing temperature towards the glass transition temperature T_g. This behavior is often interpreted in terms of a concept of "Cooperatively Rearranging Region (CRR)" [102]. It is believed that in glass-forming materials there are CRRs in which molecules must move cooperatively. With decreasing temperature towards T_g, CRRs increase in size and number, resulting in an increase of the apparent activation energy.

At a temperature slightly higher than T_g, the so-called Johari-Goldstein (JG) process is decoupled from the α-process [82]. The temperature dependence of the relaxation time of the JG process τ_{JG} is described by the Arrhenius equation in contrast to the α-process. In the polymer field, the JG process is historically called a β-process. In this article, however, in order to avoid the confusion, it is termed the JG process. As shown by Johari and Goldstein [103, 104], this JG process is commonly observed in all glass-forming materials so far investigated. Hence, it is hard to assign this process to a relaxation process of some particular groups of molecules. At the present stage, molecular origin is not elucidated [105].

It is now considered that the three processes – the α-process, the JG process, and the fast process – are commonly observed in most glass-forming materials including polymers and small molecules. In fact, the relaxation time map of *ortho*-terphenyl (OPT) [84], a typical low molecular weight glass-forming material, is very similar to that of PB. Therefore, these three processes should not be considered as special features of polymers but as common features of glass-forming materials. In what follows we discuss how these three processes are studied using neutron scattering.

4.1.2
α-Process

The α-process directly governs viscosity of the glass-forming polymers in the supercooled state, which drastically increases more than several orders of magnitude with decreasing temperature towards T_g. Hence, it is necessary to employ many kinds of experimental techniques to cover the very wide frequency region. Neutron scattering techniques can cover a frequency range faster than $\sim 10^8$ Hz, meaning that we observe rather fast motions in glass forming polymers.

Richter and coworkers [80] have measured normalized intermediate scattering functions $I(Q,t)/I(Q,0)$ from the α-process of PB using a neutron spin echo technique in a temperature range between 200 K and 280 K above the glass transition temperature (T_g=181 K). These measurements were performed at a first maximum position Q_m (=1.48 Å^{-1}) of the structure factor $S(Q)$. They found that the intermediate scattering function $I(Q, t)$ is not described by a single exponential function but by a stretched exponential function or KWW function:

$$I(Q,t) = \exp[-(t/\tau_\alpha)^\beta] \quad (0 < \beta < 1) \tag{24}$$

This anomalous relaxation behavior is often observed in complex systems including glass-forming materials and interpreted in terms of a wide distribution of the relaxation time. Using the distribution function $g(\ln \tau_\alpha)$, the observed stretched exponential function can be described by

$$I(Q,t) = \int_{-\infty}^{\infty} g(\ln \tau_\alpha) \exp(-t/\tau_\alpha) d(\ln \tau_\alpha) \tag{25}$$

It is believed that the wide distribution of the relaxation time must be due to the heterogeneity [106] (see Sect. 6). It was also found [80] that the intermediate scattering functions $I(Q, t)$ can be scaled to a master curve using shift factors a_T determined from the temperature dependence of the viscosity as shown in Fig. 10 (upper curve). This result suggests that the neutron scattering can see the same α-process as that observed in macroscopic viscosity measurements and the relaxation mechanism does not change in the temperature range examined.

They also evaluated a fraction of the α-process in $I(Q, t)$, which corresponds to the non-ergodic parameter f_ε in MCT, as a function of temperature in the region below and above T_g. According to the MCT f_ε is given by

$$f_\varepsilon(Q,T) = f_0(Q) + \begin{cases} f(Q)\sqrt{\varepsilon} & (T < T_c) \\ O(\varepsilon) & (T > T_c) \end{cases} \tag{26}$$

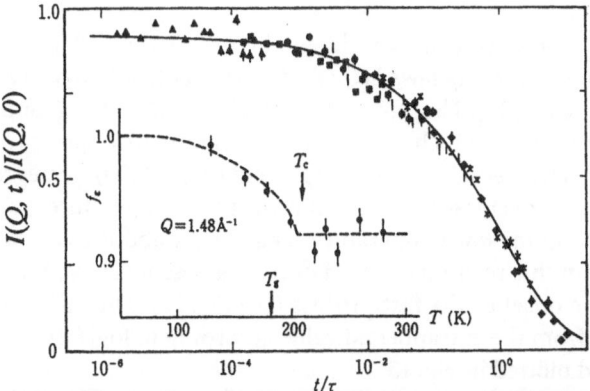

Fig. 10. Intermediate scattering functions $I(Q,t)/I(Q,0)$ of polybutadiene (PB) obtained by neutron spin echo measurements, which are scaled to a master curve using a shift factor a_T evaluated from the viscosity. *Inset* shows the temperature dependence of the non-ergodic factor, which is a fraction of the α-process. (Reprinted with permission from [80]. Copyright 1988 American Physical Society, New York)

where T_c is a critical temperature in MCT and $\varepsilon = T - T_c$. The observed temperature dependence agrees with the theoretical prediction and the critical temperature T_c estimated is about 220 K, which is ~30 K above T_g (Fig. 10). This is also consistent with the prediction. It was considered that this experiment is one direct proof of the MCT, but a vast amount of discussion on the MCT still takes place. Discussion on the MCT is out of the scope of this reviewer. Readers interested in such discussion can consult the literature. [7, 72, 107].

4.1.3
Johari-Goldstein Process

Richter et al. carried out neutron spin echo measurements at the minimum position Q_{min} of $S(Q)$ on the same polymer (PB) as that described above [82]. The intermediate scattering functions were described by a stretched exponential function as well, but could not be scaled to a master curve using a shift factor a_T. The relaxation times extracted from the observed stretched exponential functions are plotted in the relaxation time map in Fig. 9, from which it is seen that they deviate from the relaxation time of the α-process and the temperature dependence of τ_{JG} is well described by the Arrhenius formula. It was confirmed that the process observed in PB at the minimum position Q_{min} in $S(Q)$ is the JG process. The fact that the JG process is observed at Q_{min} suggests that the process is not a cooperative motion but an isolated one.

4.2
Fast Process

4.2.1
Detection of the Fast Process

First we show how we can experimentally detect the fast process. Figure 11 shows dynamic scattering laws $S(Q,\omega)$ of cis-1,4-polybutadiene (PB) below and above T_g (=170 K) [32]. They are obtained with an inverted geometry type of a neutron spectrometer, which is the cause for asymmetric spectra about the origin (ω=0). The elastic scattering intensity $I_{el}(Q)$ integrated within $0 \leq \hbar\omega \leq 0.2$ meV decreases with increasing temperature according to $\log[I_{el}(Q)] \alpha - T$ in the low temperature range below about 120 K, while it begins to deviate from this relationship and decreases steeply above this temperature, suggesting the onset of the fast process (see Fig. 12). The difference of the observed $I_{el}(Q)$ from the extrapolated value according to $\log[I_{el}(Q)] \alpha - T$ is defined as $\Delta I_{el}(Q)$ and plotted in Fig. 13.

The mean square displacement $<u^2>$ evaluated from the Q dependence of $I_{el}(Q)$ is proportional to T below about 120 K, but above about 120 K it begins to exceed the value expected from the linear relationship $<u^2> \alpha T$. This also suggests the onset of the fast process (see Fig. 14). The excess value $\Delta<u^2>$ is also plotted in Fig. 13.

In the dynamic scattering law a broad boson peak is observed at ~2.5 meV for PB and the intensity increases with temperature according to the Bose factor at low temperature far below T_g as mentioned in Sect. 3. The dashed lines in Fig. 11 are the expected value calculated from the Bose factor. At low temperatures below 120 K, the agreement between the observed and expected values is very good. On the other hand, the dynamic scattering law $S(Q,\omega)$ at 170 K ($=T_g$) shows excess scattering intensity $\Delta I_{qes}(Q,\omega)$ below ca. 4 meV compared with the expected value and the excess intensity increases with temperature as can be seen in the spectra at higher temperatures. The excess intensity $\Delta I_{qes}(Q,\omega)$ corresponds to the fast process as expected from the temperature dependence of $I_{el}(Q)$ and $<u^2>$. $\Delta I_{qes}(Q,\omega)$ is also plotted in Fig. 13.

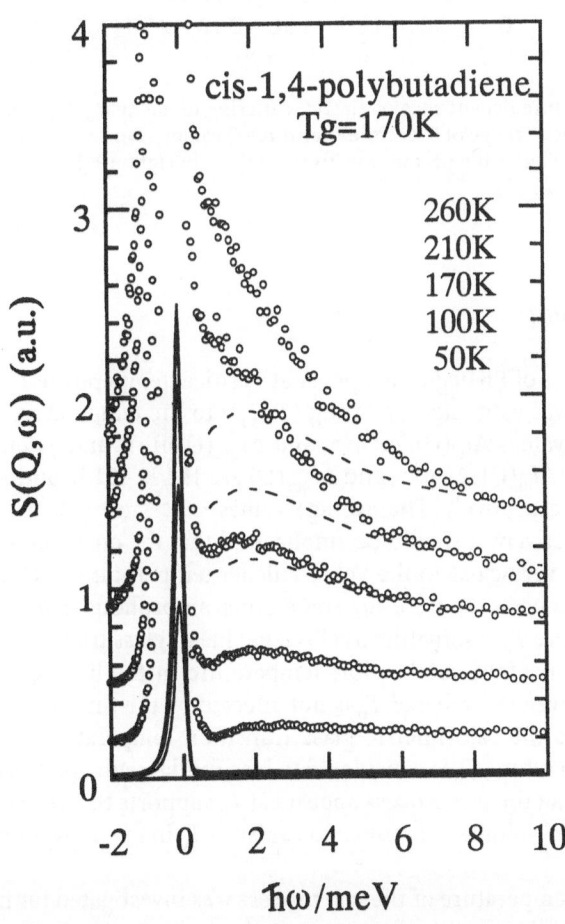

Fig. 11. Dynamic scattering law $S(Q,\omega)$ of *cis*-1,4-polybutadiene (PB) below and above the glass transition temperature T_g=170 K. *Dashed lines* are the values expected from the Bose factor. (Reprinted with permission from [32]. Copyright 1993 American Institute of Physics, New York)

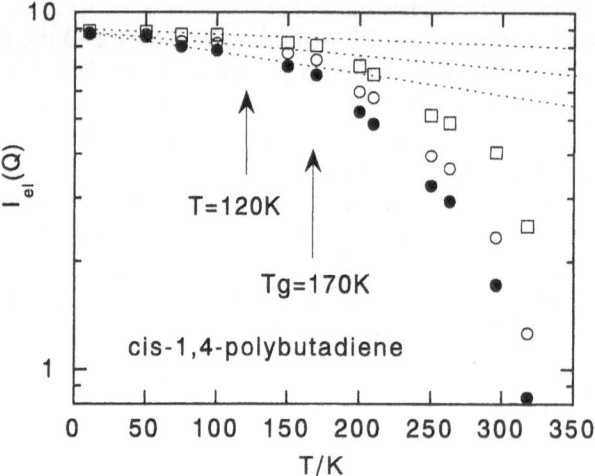

Fig. 12. Temperature dependence of elastic scattering intensity $I_{el}(Q)$ obtained by integrating $S(Q,\omega)$ in the ω-range of 0–0.2. meV for three Q values. (■): 0.62 Å^{-1}, (○): 1.76 Å^{-1}, (●): 2.35 Å^{-1}. (Reprinted with permission from [32]. Copyright 1993 American Institute of Physics, New York)

4.2.2
Onset Temperature

The fast process of PB begins to appear at a critical temperature T_f below T_g. This temperature was estimated by fitting $(T-T_f)^\gamma$ to the temperature dependencies of the excess values $\Delta I_{el}(Q)$, $\Delta <u^2>$, and $\Delta I_{qes}(Q,\omega)$. T_f and γ determined from this fitting for $\Delta I_{el}(Q)$, $\Delta <u^2>$, and $\Delta I_{qes}(Q)$ are 124 K, 121 K, and 116 K, and 1.7, 1.9, and 1.8, respectively. The average values $<T_f>$ and $<\gamma>$ are 120±7 K and 1.8±0.3, respectively. It should be emphasized that the onset temperature of the fast motion is very close to the Vogel-Fulcher temperature T_0 [101]. Similar results have also been reported for some other amorphous materials [108, 109]. The temperature T_0 is sometimes called the "ideal" glass transition temperature, i.e., an experimentally inaccessible temperature in the limit of infinitely slow cooling. However, the value of T_0 is not affected by the time scale of the experiment, whereas the calorimetric glass transition temperature depends on the time scale. Therefore, T_0 is considered to be a fundamental property of the sample. The fact that the fast process appears at T_0 supports the free volume concept because the free volume is assumed to vanish at T_0 in the derivation of the Vogel-Fulcher law.

The onset temperature of the fast process was investigated for many kinds of polymers, revealing that the onset temperatures are not always the Vogel-Fulcher temperature. In Fig. 15, for example, dynamic scattering laws of atactic polystyrene (PS) [94] are shown in a temperature range well below T_g (=373 K). At a very low temperature such as 21 K, the boson peak is clearly observed while

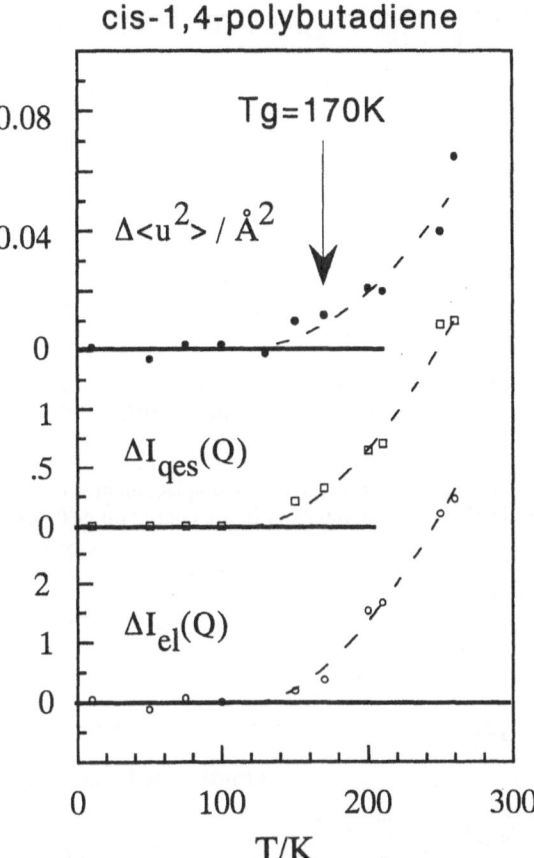

Fig. 13. Excess decrease of the elastic scattering intensity $\Delta I_{el}(Q)$ (O) at Q=1.76 Å^{-1}, excess quasielastic scattering intensity $\Delta I_{qes}(Q)$ ($\square$) at Q=1.76 Å^{-1}, and excess mean square displacement $\Delta<u^2>$ (●) as a function of temperature. The *dashed lines* are the results of the fit with $(T-T_f)^{\gamma}$. Critical temperature T_f and critical exponent γ determined in the fit for $\Delta I_{el}(Q)$, $\Delta<u^2>$ and $\Delta I_{qes}(Q)$ are 124 K, 121 K, and 116 K, and 1.7, 1.9, and 1.8, respectively. Average values $<T_f>$ and $<\gamma>$ are 120±7 K and 1.8±0.3, respectively. (Reprinted with permission from [32]. Copyright 1993 American Institute of Physics, New York)

above about 200 K or 175 K below T_g the spectra become quasielastic-like. The elastic scattering intensity $I_{el}(Q)$ measured with an energy resolution of 0.02 meV also shows a steep decrease above about 200 K. These facts suggest that the fast process appears above 200 K in PS. This onset temperature is far below the Vogel-Fulcher temperature T_0 (=325 K). In the case of PB, the fast process is attributed to the backbone chain motion because PB has no side groups. On the other hand, PS has a side group (phenyl ring) and it is expected that relaxation of the side group can occur far below T_g. In fact, Yano and Wada [110] measured mechanical relaxation of PS below T_g and found γ-relaxation at ca.

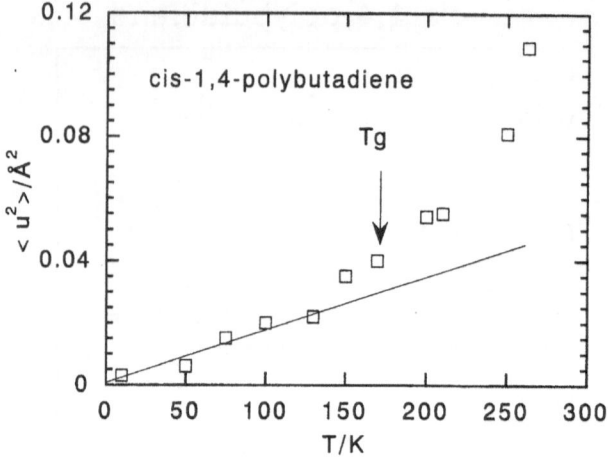

Fig. 14. Temperature dependence of mean square displacement $<u^2>$ evaluated from the Q dependence of the elastic scattering intensity in the energy range 0–0.2 meV

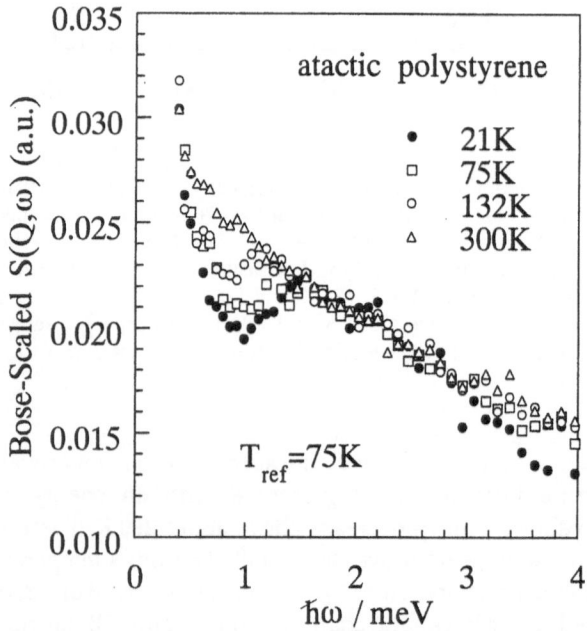

Fig. 15. Inelastic scattering intensity of atactic polystyrene below 300 K scaled by the Bose factor. The reference temperature is 75 K. (Reprinted with permission from [94]. Copyright 1996 American Institute of Physics, New York)

200 K, which was assigned to rotational motion of phenyl rings around the bond connected to the backbone chain. This temperature for the γ-relaxation is very close to the onset temperature of the fast motion in PS. It can, therefore, be considered that the change of the spectra from inelastic-like to quasielastic-like far below T_g is caused by the relaxation motion of the phenyl ring rotation. This idea has been directly confirmed in an inelastic neutron scattering experiment on PS with deuterated phenyl group [94].

Surveying the published data, the fast process appears at the Vogel-Fulcher temperature for simple polymers which have no large side groups such as *cis-trans*-polybutadiene [92], *trans*-1,4-polychloroprene (PCP) [35], polyisobutylene (PIB) [89], and atactic polypropyrene (a-PP) [95]. On the other hand, it appears far below T_g for polymers having large side groups or large internal degree of freedoms such as PS [94], PMMA [111], polyimide [39], and polycarbonate [91]. These results confirm the idea that onset of the fast process far below T_g must be due to motions of large side groups and/or internal degrees of freedom. Recently an empirical relation between the onset temperature and carbon-carbon torsional barrier in polymer chains was discussed by Colmenero and Arbe [112]

4.2.3
Characteristic Properties and Origin of the Fast Process

Generally speaking, there are two interpretations for the fast process. One is a relaxation process and the other is a softening of the vibrational states. At present we do not have a final conclusion and hence in this section we show both the interpretations using the data of a polyimide [39] as an example. Dynamic scattering law $S(Q,\omega)$ of the polyimide is shown in Fig. 16 in a temperature range below T_g (=473 K). Dashed lines in the figure are the expected value from the Bose factor. Excess quasielastic intensity is clearly observed above about 200 K, showing the onset of the fast process at a very low temperature below T_g.

In order to see the characteristic features of the fast process, the dynamic scattering law $S(Q,\omega)$ was first analyzed using a very simple phenomenological model, assuming that the vibrational density of states $G(\omega)$ is temperature independent, i.e., no softening of the vibrational modes occurs in the temperature range examined. In this case, the excess scattering intensity due to the fast process can be extracted simply by subtracting the vibrational contributions predicted from the Bose factor. It is further assumed that the excess scattering is described with a Lorentzian function. Then, the incoherent dynamic scattering law is given by

$$S(Q,\omega) = \frac{n_s(T)}{n_s(T_s)} S(Q,\omega)_{T=100K} + A(Q)L(Q,\omega,\Gamma) \qquad (27)$$

where $n_s(T)$ and $S(Q,\omega)_{T=Ts}$ are the Bose factor and the scattering law at a standard temperature T_s, respectively. $L(Q,\omega,\Gamma)$ is a Lorentzian function with half-width at half-maximum Γ, the inverse of Γ corresponds to a characteristic time

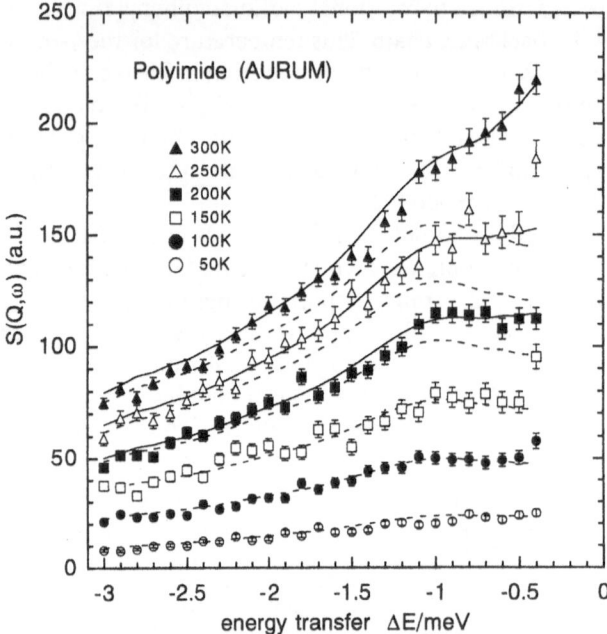

Fig. 16. Dynamic scattering law $S(Q,\omega)$ of the polyimide at $Q=2.5\ \text{Å}^{-1}$ as a function of temperature. *Dashed lines* are the values expected from the Bose factor. *Solid lines* are the results of the fit by a simple phenomenological model assuming that the softening of vibrational modes does not occur (see text). (Reprinted with permission from [39]. Copyright 1998 American Institute of Physics, New York)

of the fast process τ_f $(=h/\Gamma)$, and $A(Q)$ is its weight. The results of the fits with the model are shown by solid curves in Fig. 16. The agreement is very good, suggesting that a single exponential function describes the excess scattering due to the fast process well. The evaluated half-width Γ $(=h/\tau_f)$ is plotted in Fig. 17, indicating that Γ is almost independent of temperature. Similar results have been reported for many amorphous polymers as well as molecular glass-forming materials. The fact that τ_f is independent of T suggests that the fast process is not a motion passing over an energy barrier but a localized motion in a confined region or in a potential well. This is also confirmed by a fact that Γ is independent of Q. Taking into account that the spectrum of the fast process is quasielastic-like, two candidates for the fast process were proposed. One is an over-damped vibrational motion in a potential well and the other is a relaxational motion between two or more quasi-equilibrium sites separated by a very low energy barrier. In the case of an over-damped vibration, the potential must be of an up-ended bell-shape because excess value of the mean square displacement $<u^2>$ increases with increasing temperature (see Fig. 14), which is schematically displayed in Fig. 18a. The second candidate, a relaxational motion between quasi-equilibrium sites, can be expected in disordered systems where there exist many

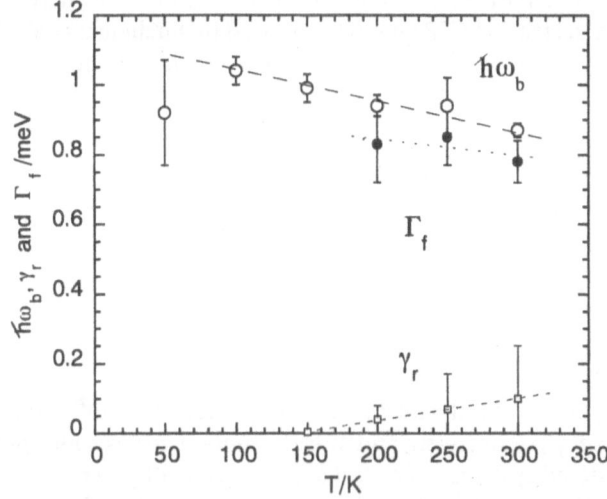

Fig. 17. Relaxation rate $\Gamma (=h/\tau_f)$ ($\bullet$) of the fast process evaluated from a simple phenomenological model (Fig. 16) and the boson peak frequency ω_b (O) and the friction term γ ($\square$) evaluated from the VR model (Fig. 19).(Reprinted with permission from [39]. Copyright 1998 American Institute of Physics, New York)

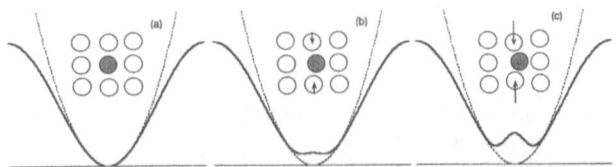

Fig. 18a–c. Schematic sketch of up-ended bell-shape potential models: **a** without; **b, c** with potential softening due to different strain from surrounding atoms in disordered systems. *Closed and open circles* are schematic representations of marked and surrounding atoms (or molecules).(Reprinted with permission from [39]. Copyright 1998 American Institute of Physics, New York)

low energy barriers produced by the strain from the surrounding atoms. It is schematically demonstrated for two cases with different strains in Fig. 18b,c. The concept of the potential softening in disordered systems has been extensively discussed in the soft-potential (SP) model [68], which could explain the low temperature behavior in disordered systems such as vibrational density of states, heat capacity, and thermal conductivity as mentioned in Sect. 3.

In the model, it is assumed that no softening of vibrational modes occurs with increasing temperature. In an inelastic neutron scattering experiment on polyisobutylene (PIB) [33] it was found that the boson peak is observable even above the glass transition temperature T_g and sustains softening as temperature increases. Taking into account such softening with increasing temperature as well as relaxational process, the data of the polyimide were analyzed in terms of the vibration-relaxation (VR) model proposed by Buchenau et al. [91, 92] to see the possibility of vibrational softening. In the VR model, the scattering law due to the vibrational contribution is given by an empirical function:

$$S_{\text{vib}}(Q,\omega) = \frac{3k_B TQ^2}{2M\omega_0}\left(\frac{A+\omega^2}{B+\omega^4}\right)^{1/2} \tag{28}$$

$$A = \frac{\omega_0^4\omega_b^4}{\omega_D^6 - 2\omega_0^4\omega_b^2} \qquad B = \frac{\omega_D^6\omega_b^4}{\omega_D^6 - 2\omega_0^4\omega_b^2} \tag{29}$$

where ω_b, ω_0, and ω_D are the boson peak frequency, the cutoff frequency and the Debye frequency, respectively, and M is the mass. The basic idea of the model is that in disordered systems atoms feel a wide variety of potential fields and in an extreme case they might feel negative force constant when they are on the top of the potential barrier. Excitations in such a situation was described by a random dynamic matrix method of Wigner [113].

The relaxational part of the scattering law in a symmetric double well potentials is calculated assuming that the anharmonic interaction between different modes acts like stochastic forces on a given mode, giving rise to a Langevin equation with a friction term γ_r. The scattering law $S_{\text{relax}}(Q,\omega)$ due to the relaxational mode is given by

$$S_{\text{relax}}(Q,\omega) = \frac{2k_B TQ^2}{M\omega_0}\int_1^{\infty}dv\,\frac{2v\gamma_r\exp(-v)/\pi}{4v^2\gamma_r^2\exp(-2v)+\omega^2} \tag{30}$$

$$v = E_a / k_B T \tag{31}$$

The results of the fitting with the VR model are shown in Fig. 19; the fit is very good. It was found in this model that the excess scattering intensity due to the fast process can be explained mainly by the softening of the vibrational motions. The boson peak energy γ_r and the friction term γ_r are shown as a function of temperature in Fig. 17. The softening of the boson peak is clearly observed with increasing temperature. The value of the relaxation rate γ_r at each temperature is much smaller compared with the observed energy region. This means that only the high frequency tails of the relaxation spectra were observed in the present examined energy range. In other words, the observed spectra are mainly governed by the vibrational contribution, implying the important role of the softening of vibrational modes in the VR model.

Another interpretation of the fast process has been proposed on the basis of the free-volume concept presented by Cohen and Turnbull [114]. It was reported for selenium [115] and PB [116] that the inverse of the mean square displacement difference between the values in the disordered and the ordered phase

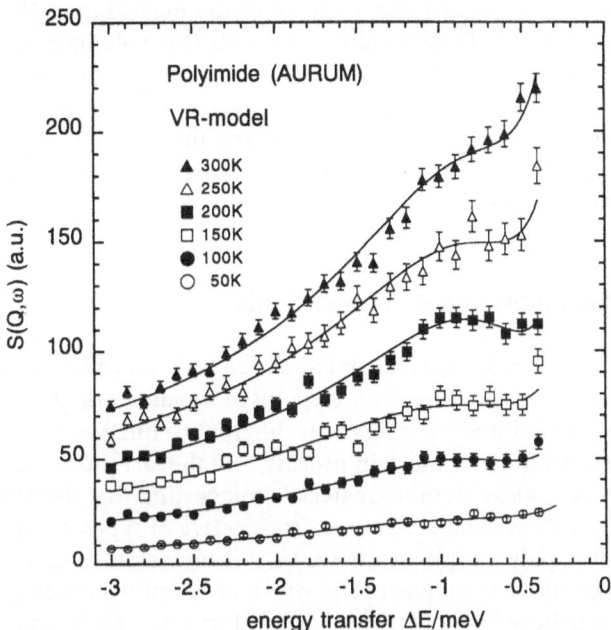

Fig. 19. Results of the fit with the VR model (*solid lines*). The data points are the same as in Fig. 16. (Reprinted with permission from [39]. Copyright 1998 American Institute of Physics, New York)

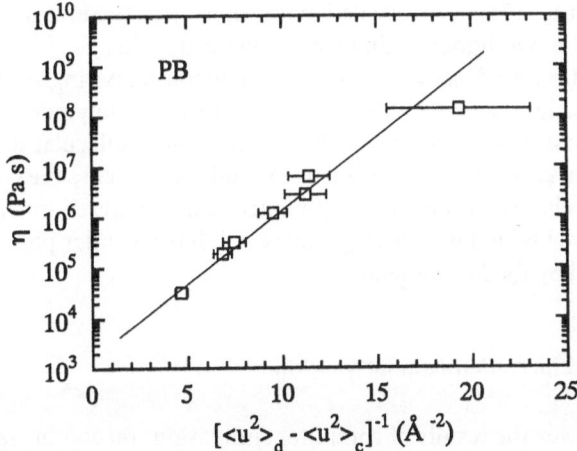

Fig. 20. Relation between logarithm of the viscosity log η and inverse of the excess mean square displacement $[<u^2>_d - <u^2>_c]^{-1}$. (Reprinted with permission from [116]. Copyright 1999 American Physical Society, New York)

$[1/(<u^2>_d - <u^2>_c]$ is linearly proportional to the logarithm of the viscosity as shown in Fig. 20 for PB. These results support the free-volume concept [114] and it was suggested that the free-volume could be replaced by the mean square displacement of the fast motion. The fact that the fast process of PB disappears at the Vogel-Fulcher temperature T_0 also supports this concept since the free-volume is assumed to vanish at T_0 in the derivation of the Vogel-Fulcher law within the free-volume concept.

5
Conformational Transition in Polymer Liquids

As discussed in Sects. 3 and 4, the dynamics of bulk amorphous polymers have been extensively studied from the viewpoint of dynamics of glass-forming materials. Experimental studies using many different techniques have revealed that the α-process, the Johari-Goldstein process, and the fast process are commonly observed in most glass-forming materials, suggesting that dynamics near the glass transition show general features, irrespective of types of materials. From the viewpoint of polymer physics, however, it is important to understand which properties are unique to polymers and which are common for all glass-forming materials. It is believed in polymer science that structural relaxation of the main chain backbone begins near the glass transition temperature T_g, meaning that the polymer chains begin to change their conformations near T_g.

Extensive theoretical and experimental works were carried out on local dynamics of polymers in solution and bulk to elucidate the mechanism of conformational transitions [106]. Formerly, it was believed that the most reasonable mechanism for the conformational transitions was a crankshaft-like motion such as the Schatzki crankshaft [117] or three-bond motions [118, 119] in which two bonds in a main chain rotate simultaneously. However, recent computer simulations [120–128] have revealed many interesting features of conformational transitions of a polymer chain in solutions and melts.

In this section we describe quasielastic neutron scattering studies [129, 130] focusing attentions on conformational transitions of polymer chains. For this purpose we first summarize the results of the recent molecular dynamics (MD) simulations on conformational transitions and then discuss the conformational transition mechanism on the basis of neutron data analyzed in terms of a jump diffusion model with damped vibration which has a similar physical picture to that predicted by the MD simulations.

5.1
MD Simulations on Conformational Transition

We briefly survey the results of computer simulations on conformational transitions in polymer chains. A pioneering work was made on an isolated polyethylene-like model by Helfand and coworkers [120, 121] on the basis of the Langevin equation to reveal that single bond rotations are possible in a polymer chain for conformational transitions as well as cooperative rotations. The latter is mainly manifested in the counter rotations of two second neighboring bonds separated by a *trans* bond. The key to understanding the conformational transitions is that distortions in a polymer chain due to the strain brought about by the single bond rotation can be relaxed through deformations of degrees of freedom in the neighborhood of the conformation-transforming bond. It was also elucidated by the simulations that the relaxation time of the elementary process for the conformational transitions is in the range of several tens to several hundreds of picoseconds and the apparent activation energy of a conformational transition is approximately equal to the energy barrier of a single C-C bond rotation.

Following the work of Helfand and coworkers, MD simulations on bulk amorphous polymers were performed by several groups; on a polyethylene-like model by Takeuchi and Roe [122, 123] and by Boyd and coworkers [124], and on *cis*- and *trans*-polybutadienes by Gee and Boyd [125] and by Kim and Mattice [126]. These MD simulations showed that the motions of a polymer chain in the long time range are affected by surrounding molecules acting as walls, while in the short time range the motions or the mechanisms of conformational transitions are surprisingly similar to those observed in an isolated chain by Helfand and coworkers, i.e., single bond rotations as well as counter rotations of two second neighboring bonds are possible in a polymer chain. Such conformational transitions can be localized by deformations of other degrees of freedom of neighboring bonds and the localization is extended up to four or five bonds. In other words, because of many degrees of freedom, polymer chains are so flexible to make the conformational transition localized. Furthermore, it was shown that even in bulk the apparent activation energy of the elementary process for the conformational transitions is roughly the same as the energy barrier height of a single C-C bond rotation. This is the same result as for an isolated chain.

5.2
Onset of E-Process

According to the computer simulations, local conformational transitions must be observed in a time range of several tens of picoseconds to several hundreds of picoseconds, which correspond to an energy region of several hundreds to several tens of μeV. To confirm if such a dynamic process is observed in $S(Q,\omega)$, quasielastic neutron scattering measurements were performed with an inverted geometry type neutron spectrometer on *cis*-1,4-polybutadiene (PB) [130] in an energy range from 0.016 meV to 0.5 meV above T_g. The observed spectrum at 260 K is shown in Fig. 21 where a spectrum observed in an energy region from 0.2 meV to 2 meV is connected. The spectrum consists of at least two quasielastic components, broad and narrow; the broad component is the fast process having an energy of 1~2 meV which has been discussed in Sect. 4 and the narrow component is a new slow process having an energy of ~100 μeV, which is termed E-process hereafter. In the inset of Fig. 21, integrated intensities of the fast and the E- processes are plotted against temperature. As mentioned in Sect. 4, the

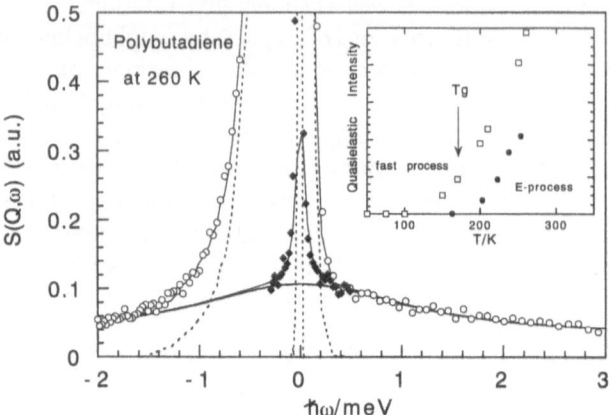

Fig. 21. Dynamic scattering law $S(Q,\omega)$ of polybutadiene at 260 K measured by two neutron TOF spectrometers LAM-40 and LAM-80 with energy resolutions $\delta\varepsilon$=0.2 meV (O) and 0.016 meV (◆). *Dashed lines* are the resolution functions of LAM-40 and LAM-80. *Inset* represents temperature dependence of integrated intensity of the fast process (□) and the E-process (●). (Reprinted with permission from [130]. Copyright 1999 American Chemical Society, Washington)

fast process of PB appears at around the Vogel-Fulcher temperature T_0 while the new E-process appears at around the glass transition temperature. It was also found that the spectra of the E-process are well described by single exponential functions. A similar E-process is also observed in polyethylene (PE), polychloroprene (PCP), and polyisobutylene (PIB).

5.3
E-Process in Relaxation Time Map

In order to identify the E-process of PB, the characteristic times are plotted in a relaxation time map (Fig. 22) [130]. As mentioned in Sect. 4, the three processes – the α-process, the Johari-Goldstein process, and the fast process – are commonly observed in most glass-forming materials including polymers and small molecules. Therefore, these three processes should not be considered as special features of polymers but as common features of glass-forming materials. On the other hand, the E-process is not observed in the relaxation time map of OTP [84], suggesting that the E-process is characteristic of polymers.

The relaxation times of the E-process are intermediate between the fast process and the α-process below about 320 K, while they almost agree with those of the α-process above ~320 K. The activation energy of the E-process, which was evaluated above ~300 K, is 2.5 kcal/mol (see Fig. 23). This value corresponds to the barrier height of a single C-C bond rotation. Similarly, the activation energies of the E-process for PCP and PIB were obtained as 2.6 kcal/mol and 2.9 kcal/mol, respectively. These values are close to the activation energy pre-

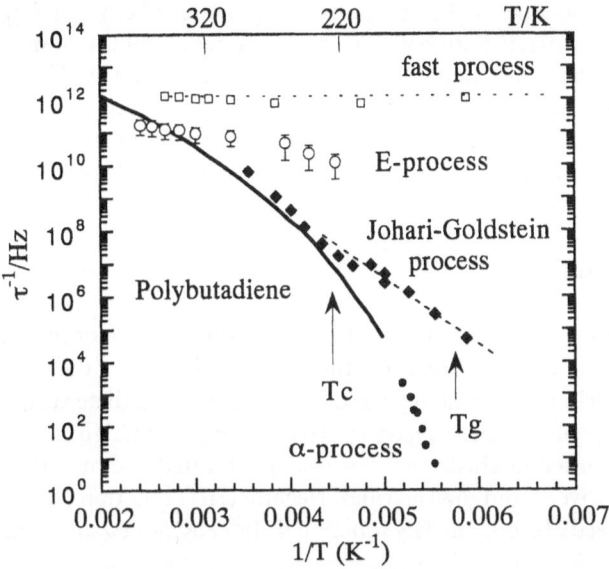

Fig. 22. Relaxation time map of polybutadiene observed using various experimental methods (see text). QENS (LAM-40) (□) [79, 129], QENS (LAM-80) (○) [129, 130], NSE (◆) [81, 82], viscosity (———) [100], ²H NMR (●) [99]. (Reprinted with permission from [130]. Copyright 1999 American Chemical Society, Washington)

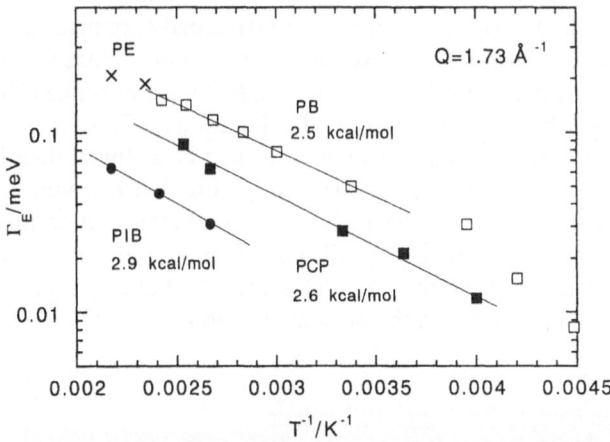

Fig. 23. Temperature dependence of the half-width at half-maximum (HWHM) Γ of the Lorentzian function fitted to the E-process at $Q=1.73$ Å^{-1}. Numerical values in the figure are apparent activation energies from the straight lines. (□): PB, (●): PIB, (■): PCP, (×): PE. (Reprinted with permission from [130]. Copyright 1999 American Chemical Society, Washington)

dicted for conformational transitions in a polymer chain by computer simulations. Furthermore, the observed relaxation times are of the same order as those predicted by the simulations, and the spatial scale probed by Q in the experiment is of the order of that expected for conformational transitions. These features strongly suggest that the E-process is an elementary process for local conformational transitions. This is the reason why it is called the E-process.

5.4
Tail of the α-Process?

It is well known that the α-process has a wide distribution of the relaxation time and hence there is a possibility that the E-process is a tail of the α-process. Due to the wide distribution of the relaxation time the intermediate scattering function due to the α-process is described by a stretched exponential function [exp(-$(t/\tau)^\beta$); $0<\beta<1$]. In order to check this possibility, we fitted a dynamic scattering law, $S_{HN}(Q,\omega)$, derived from the Havriliak-Negami (HN) function (see Eqs. 32 and 33) to the observed $S(Q,\omega)$. The HN function $\Phi^*(\omega)$ and $S_{HN}(Q,\omega)$ are given by

$$\Phi^*(\omega) = \frac{1}{[1+(i\omega\tau_{HN})^\alpha]^\gamma} \tag{32}$$

$$S_{HN}(Q,\omega) \propto -\frac{1}{\omega}\text{Im}[\Phi^*(\omega)] \tag{33}$$

where α and γ are two parameters in the range ($0<\alpha$, $\gamma<1$) and τ_{HN} is a characteristic time of the relaxation process. According to Alvarez et al. [131, 132], the dynamic scattering law $S_{HN}(Q,\omega)$ is identical to the numerically calculated scattering law for the Fourier transform of the stretched exponential function, and the exponent β in the stretched exponential function is close to the product of the α and γ parameters in the HN function, $\beta \approx \alpha\gamma$. An example of the HN fit is shown in Fig. 24a. For comparison, the Lorentzian fit is also displayed in Fig. 24b. As seen in Fig. 24a,b, the quality of the fits is almost identical whether a Lorentzian or an HN function is used, implying that it would be impossible only from the fits to distinguish which function is better to describe the spectra. If the E-process is a high frequency tail of the α-process, the present analysis corresponds to an assumption that the high frequency tail is governed by the local conformational transitions in the present Q region.

5.5
Jump Diffusion Model with Damped Vibrations

In the following an outline of the model and some of the assumptions used in the analysis are described [129]. The schematic sketch of the model is given in Fig. 25. In the model, conformational transitions are represented by a jump motion from a conformation (a rotational isomeric state) to another. In each rota-

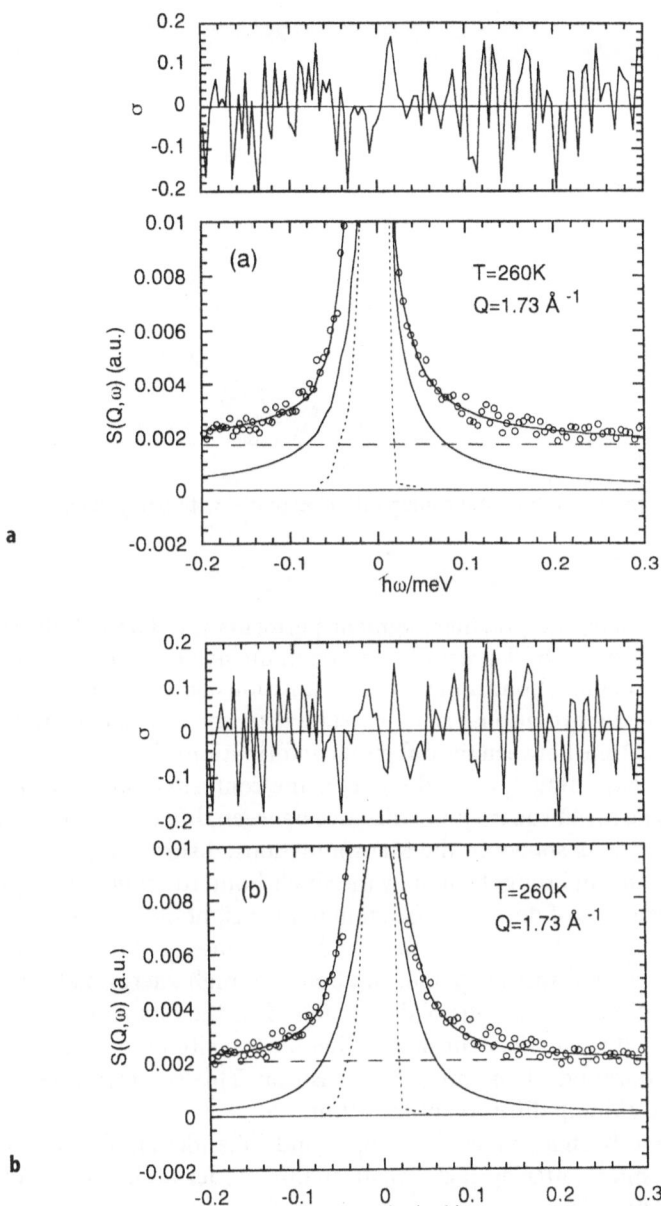

Fig. 24. a Result of fit to LAM-80 spectrum of PB at 260 K by sum of dynamic scattering law $S_{HN}(Q,\omega)$ derived from the Havriliak-Negami function (Eqs. 32 and 33) (——), δ-function (..........), and flat component (– – –). $1/\tau_{HN}$, α and γ estimated in the fit are 0.033 meV, 0.66, and 0.90, respectively. (Reprinted with permission from [130]. Copyright 1999 American Chemical Society, Washington). **b** Result of fit to LAM-80 spectrum of PB at 260 K by sum of a single exponential function (——), δ-function (..........), and flat component (– – –). $1/\tau$ estimated in the fit is 0.027 meV. (Reprinted with permission from [130]. Copyright 1999 American Chemical Society, Washington)

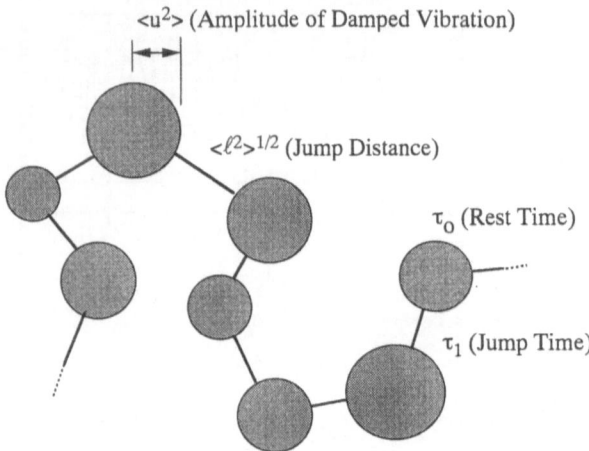

Fig. 25. Schematic representation of the jump diffusion model with damped vibrations

tional isomeric state, the polymer segment performs the damped vibrations, which assist distortions of degrees of freedom in the neighborhood of the conformation-transforming bond and keep the transition coordinate localized. The average rest time in a rotational isomeric state (lifetime of conformation) and the average jump time between the different conformations are defined as τ_0 and τ_1, respectively. Assuming $\tau_1 \ll \tau_0$, the rate of the conformational transition is given by τ_0^{-1}. The mean square jump distance between different conformations is defined as $<\ell^2>$. As shown in the MD simulations, distortions in a polymer chain due to the strain brought about by the single bond rotation can be relaxed through deformations of degrees of freedom in the neighborhood of conformation-transforming bond. It suggests that the average energy of the neighboring bonds increases due to the distortion, and bonds in high energy states tend to cause conformational transitions. This was found in the MD simulations as a counter rotation. It is therefore considered that the transitions occur successively in the neighborhood of the previous transition. This succession leads to a physical picture of jump diffusion of transition sites.

This model has been formulated by Singwi and Sjölander [133]. According to them, the dynamic scattering law $S(Q,\omega)$ of this model under the condition $\tau_1 \ll \tau_0$ is given by

$$S_{inc}(Q,\omega) = (\frac{\Gamma}{\pi})\frac{\exp[-<u^2>Q^2]}{\omega^2 + \Gamma^2} \tag{34}$$

The dynamic scattering law $S(Q,\omega)$ is a Lorentzian, which agrees with our observation and the half-width at half-maximum (HWHM) Γ of the Lorentzian function is given by

$$\Gamma = \frac{Q^2 D + (1 - \exp[-Q^2 <u^2>]) / \tau_0}{1 + Q^2 D \tau_0}$$

(35)

where D is the diffusion coefficient of the transition site defined as $D = <\ell^2>/6\tau_0$.

5.6
Mechanism of Conformational Transition

The jump diffusion model with damped vibrations was applied to the analysis of the dynamic scattering laws $S(Q,\omega)$ of the E-process far above the glass transition temperature T_g, where the conformational transitions are mainly dominated by the intramolecular interactions in a polymer chain, so that it is not necessary to take into accounts intermolecular cooperativity effects on the conformational transitions.

The theoretical equation of the jump diffusion model (Eq. 35) was fitted to the observed Q^2 dependence of HWHM Γ of the Lorentzian function of the E-process. The results are shown in Fig. 26 for PB, PIB, PCP and PE. The agreements are very good, suggesting that the model well describes the E-process. In the fit the root mean square jump distance $<\ell^2>^{1/2}$ and the relaxation time of the ele-

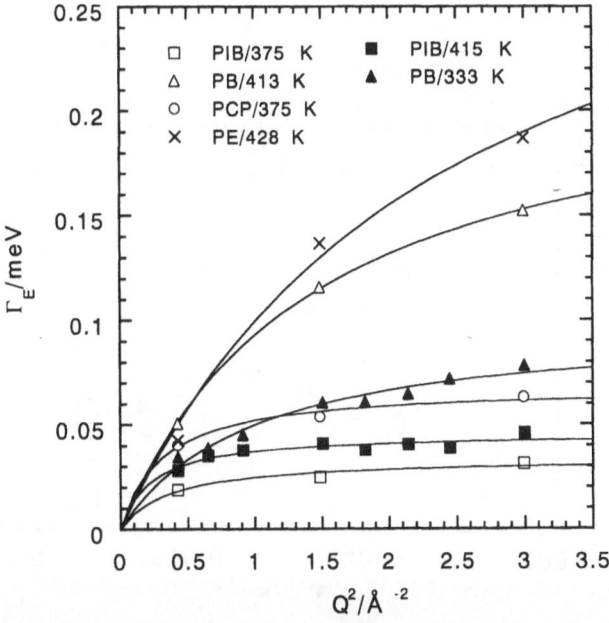

Fig. 26. Q^2 dependence of the half-width at half-maximum (HWHM) Γ of the Lorentzian function fitted to the E-process. ($\square$): PIB at 375 K, ($\blacksquare$): PIB at 415 K, ($\triangle$): PB at 413 K, ($\blacktriangle$): PB at 333 K, ($\bigcirc$): PCP at 375 K, ($\times$): PE at 428 K. (Reprinted with permission from [130]. Copyright 1999 American Chemical Society, Washington)

mentary process for the conformational transitions (or the average rest time in a conformation τ_0) were evaluated. The relaxation time is plotted vs $1/T$ in Fig. 27. The temperature dependence of the relaxation time is well represented by the Arrhenius equation and the activation energies evaluated are 4.0, 3.2 and 3.1 kcal/mol for PIB, PB and PCP, respectively.

The root mean square jump distance $<\ell^2>^{1/2}$ for the conformational transition is plotted against T in Fig. 28 for PB, PE, PCP and PIB. The jump distance is independent of temperature for each polymer within experimental error, suggesting that the mechanism of the conformational transitions does not change in the temperature range examined. In the cases of PB and PE which have no side groups, the root mean square jump distance $<\ell^2>^{1/2}$ was observed to be around 2 Å. Assuming that the conformational transitions of PB are localized within three bonds the value of $<\ell^2>^{1/2}$ is evaluated to be 2.06 Å [129]. This agrees very well with the observed value. Under the same assumption, the value of $<\ell^2>^{1/2}$ was evaluated for PE to be 1.8 Å, which is also very close to the observed value. It is therefore concluded that the conformational transitions of PB and PE are localized within three bonds, at least in average. On the other hand, the $<\ell^2>^{1/2}$ for PIB and PCP are around 5 Å, being rather large compared with those of PE and PB, suggesting that the conformational transitions of PIB and PCP are not localized within three bonds. This must be due to the local stiffness of the polymer chain.

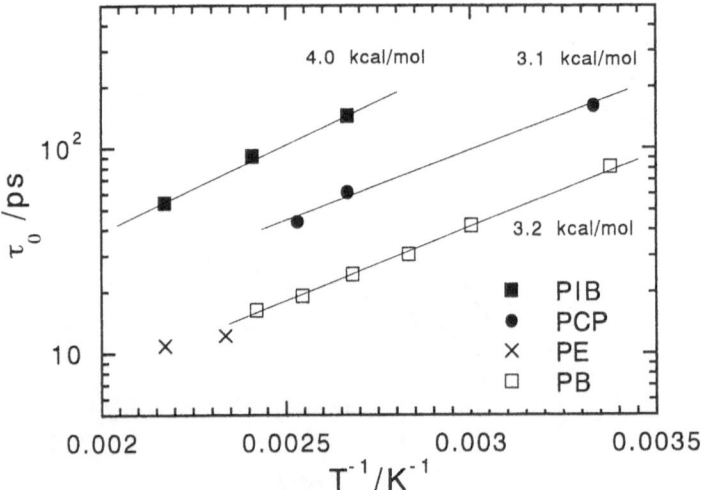

Fig. 27. Temperature dependence of relaxation time for the conformational transitions evaluated using the jump diffusion model with damped vibrations. (□): PB, (●): PIB, (■): PCP, (×): PE. (Reprinted with permission from [130]. Copyright 1999 American Chemical Society, Washington)

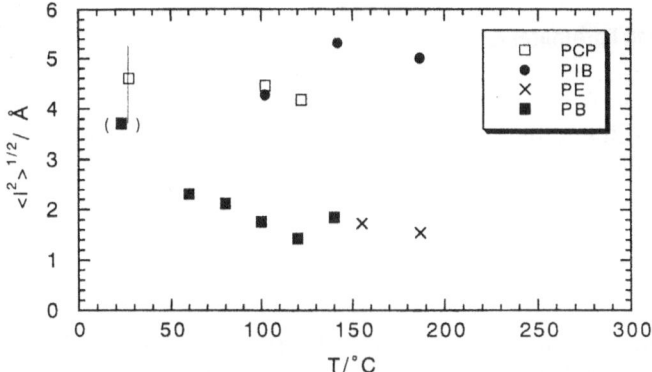

Fig. 28. Root mean square jump distance $<\ell^2>^{1/2}$ for the conformational transitions evaluated using the jump diffusion model with damped vibrations. (■): PB, (●): PIB, (□): PCP, (×): PE. (Reprinted with permission from [130]. Copyright 1999 American Chemical Society, Washington)

6
Heterogeneity of Amorphous Polymers

Recent studies on dynamics of glass-forming materials have elucidated various anomalous but common features of the dynamics near the glass transition temperature [71]. For example, density-density correlation function of the α-process does not show the Debye behavior, but it is well described by a stretched exponential function or KWW (Kohlrausch-Williams-Watts) function $\exp[-(t/\tau)^\beta]$ (0<β<1). This result is often interpreted in terms of a very broad distribution of the relaxation time of the α-process, suggesting glass-forming materials are dynamically very heterogeneous. Another anomalous feature is the temperature dependence of the relaxation time, which is not described by the Arrhenius equation but by the Vogel-Fulcher equation $\tau^{-1}=\tau^{-1}_\infty\exp[-B/(T-T_0)]$ where T_0 is a Vogel-Fulcher temperature. This anomalous temperature dependence is often interpreted by an idea of cooperativity. It is believed that there are some regions in glass-forming materials where molecules must move cooperatively [102, 134–136], which is termed cooperatively rearranging region (CRR), and the existence of the CRRs makes the glass-forming materials very heterogeneous, leading to a wide distribution of relaxation time of the α-process. Thus, many anomalous features of dynamics of glass-forming materials are explained in terms of dynamical heterogeneity. Recent NMR and fluorescence probe investigations have directly showed the dynamical heterogeneity in glass-forming materials although there are still big debates between the heterogeneous and homogeneous scenario [137]. The present situation of the studies on heterogeneity is excellently reviewed by Sillescu [138].

In this section we review recent investigations of dynamical heterogeneity of glass-forming polymers below and above T_g in terms of non-Gaussian parameter evaluated from incoherent neutron scattering data [139, 140].

6.1
Non-Gaussian Parameter A_0

Non-Gaussian parameter $A_0(t)$ was first introduced by Rahman et al. in 1962 [141]. According to them, incoherent intermediate scattering function $I(Q,t)$ is obtained from the cumulant expansion up to the order of Q^4 as

$$I(Q,t) = \exp\left(-<u^2>(t)Q^2 + \frac{1}{2}A_0(t)[<u^2>(t)]^2 Q^4\right) \qquad (36)$$

In this expression, the integrals of the velocity correlation functions, denoted by $\gamma_1(t)$ and $\gamma_2(t)$ in [141], have been expressed in terms of the mean square displacement $<u^2>$ and the non-Gaussian parameter $A_0(t)$ as

$$A_0(t) = \frac{3<u^4>(t)}{5[<u^2>(t)]^2} - 1 \qquad (37)$$

In the so-called Gaussian approximation, the intermediate scattering function is given in the form up to the order of Q^2. This approximation is perfectly held in the motions of harmonic oscillators, perfect gases and diffusion processes at infinite time though in real systems it is valid only for low Q range. In a high Q range, one has to take into account the non-Gaussian behavior. It was shown [142] that non-Gaussian parameter arises from many reasons such as dynamical heterogeneity, dynamical anisotropy, and anharmonisity. It is considered in disordered systems that the most plausible origin is dynamical heterogeneity due to local environments because the local environments are surely different for each molecule [53].

6.2
Non-Gaussian Parameter as a Measure of Heterogeneity

It is first assumed that the motion in the individual environment is Gaussian, so that the incoherent intermediate scattering function $I_{ind}(Q,t)$ is given by

$$I_{ind}(Q,t) = \exp(-<u^2>Q^2) \qquad (38)$$

It is further assumed that the mean square displacement has a Gaussian distribution $g_G(<u^2>)$

$$g_G(<u^2>) = \frac{1}{\sqrt{2\pi\sigma^2}}\exp\left\{-\frac{\left(<u^2>-\overline{<u^2>}\right)^2}{2\sigma^2}\right\} \qquad (39)$$

where $\sigma^2 = \overline{(\Delta <u^2>)^2} = \left(\overline{<u^2> - \overline{<u^2>}} \right)^2$ From Eqs. (38) and (39), the incoherent intermediate scattering function $I(Q,t)$ up to the order of Q^4 is obtained by averaging over the distribution as

$$I(Q,t) = \int_0^\infty g_G(<u^2>) \exp(-<u^2>Q^2) d(<u^2>)$$

$$\approx \exp\left(-\overline{<u^2>}Q^2 + \frac{A_0 \overline{<u^2>}^2}{2} Q^4 \right) \tag{40}$$

using the non-Gaussian parameter A_0 defined as

$$A_0 = \frac{\overline{<u^2>^2} - \overline{<u^2>}^2}{\overline{<u^2>}^2} \tag{41}$$

It is obvious that the non-Gaussian parameter is a standard deviation of the Gaussian distribution normalized to the average value, and hence it is a measure of the heterogeneity. In a frequency space, the incoherent elastic scattering intensity $S(Q,\omega=0)$ is given by

$$S(Q,\omega=0) = \exp\left(-\overline{<u^2>}Q^2 + \frac{1}{2}A_0 \overline{<u^2>}^2 Q^4 \right) \tag{42}$$

where $<u^2>$ is now the mean square displacement in frequency domain. It is noted that, in experiments, the elastic scattering intensity $I_{el}(Q)$ is measured with a limited energy resolution of a neutron spectrometer.

6.3
Temperature Dependence of Non-Gaussian Parameter

The elastic scattering intensity $I_{el}(Q)$ measured with an energy resolution $\delta\varepsilon = 1.1$ meV is shown in Fig. 29 as a function of Q^2 for polyisobutylene (PIB), where $I_{el}(Q)$ at each T is divided by $I_{el}(Q)$ at 10 K in order to reduce the effects of coherent scattering. It is clear that the $\log[I_{el}(Q)]$ is not proportional to Q^2, meaning that $S(Q,0)$ cannot be described within the Gaussian approximation. The solid lines in Fig. 29 are the results of the fits using Eq. (42), showing good agreements.

The mean square displacements $\overline{<u^2>}$ and the non-Gaussian parameters A_0 evaluated from the fits are plotted against the reduced temperature T/T_g for PB, PIB, PS, and PMMA in Fig. 30. The mean square displacements $<u^2>$ for all the

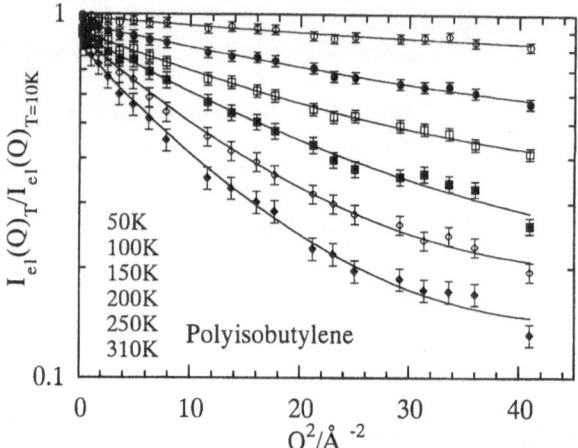

Fig. 29. Q^2-dependence of incoherent elastic scattering intensity $I_{el}(Q)_T/I_{el}(Q)_{T=10K}$ of polyisobutylene (PIB) at 50 (O), 100 (●), 150 (l), 200 (□), 250 (◇) and 300 K (◆). *Solid lines* are the results of fit to Eq. (43)

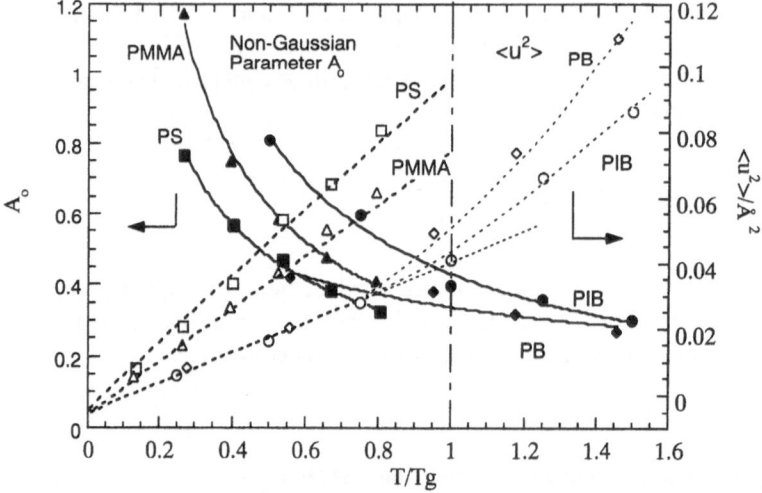

Fig. 30. Mean square displacement $<u^2>$ (*open symbols*) and non-Gaussian parameter A_o (*closed symbols*) as a function of reduced temperature T/T_g for PB (◆, ◇), PIB (●, O), PS (■, □), and PMMA (▲, △). (Reprinted with permission from [140]. Copyright 1998 Elsevier Science B. V., Amsterdam)

samples increase linearly with temperature below the glass transition temperature ($T/T_g = 1$), suggesting that the motion observed can be approximated as a harmonic motion. On the other hand, $\overline{<u^2>}$ deviates from the linear relationship above T_g, showing excess values of the mean square displacement. In the case of PB, the temperature of the deviation is slightly lower than T_g. The excess

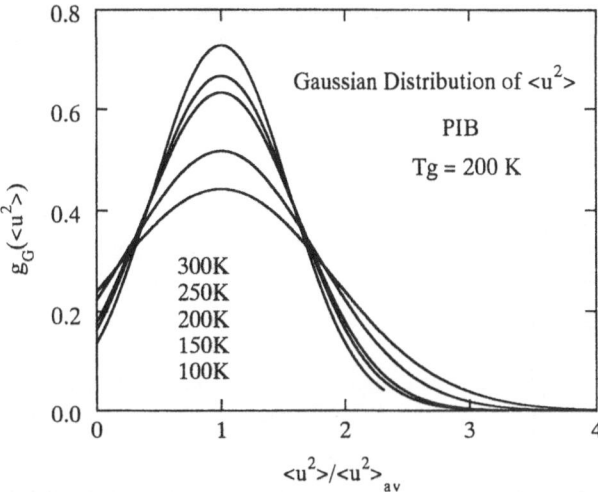

Fig. 31. Gaussian distribution function $g_G(<u^2>)$ of mean square displacement for PIB

value of $<u^2>$ indicates the onset of the so-called fast process of the order of picosecond as discussed in Sect. 4.

The non-Gaussian parameter A_0 seems to increase slightly as temperature decreases above T_g for PIB and PB, while it increases steeply below T_g for PIB, PB, PS, and PMMA. Similar temperature dependencies were observed for all the samples examined which indicates that amorphous polymers become more heterogeneous with decreasing temperature.

In Fig. 31, the evaluated Gaussian distribution functions of $<u^2>$ for PIB (Eqs. 39 and 40) are displayed at temperatures below and above T_g (=200 K), where the mean square displacement is normalized to the average value ($<u^2>_{av}$). Clearly, the distribution becomes broader with decreasing temperature.

6.4
Time Dependence of Non-Gaussian Parameter

Odagaki and Hiwatari [143, 144] developed the trapping diffusion theory to describe glass transition phenomena and showed that the glass transition is a transition from Gaussian to non-Gaussian behavior at infinite time. They calculated time dependence of the non-Gaussian parameter below and above the glass transition temperature. Even above the glass transition temperature, the Gaussian behavior is not observed until infinite time. In a short time region, above T_g, the non-Gaussian parameter A_0 increases with time, then after reaching a maximum it decreases with time to reach zero at infinite time while below T_g it has a finite value even at infinite time.

Time evolution of A_0 was experimentally examined by changing the energy resolution of the spectrometer to see the theoretical predictions. For this pur-

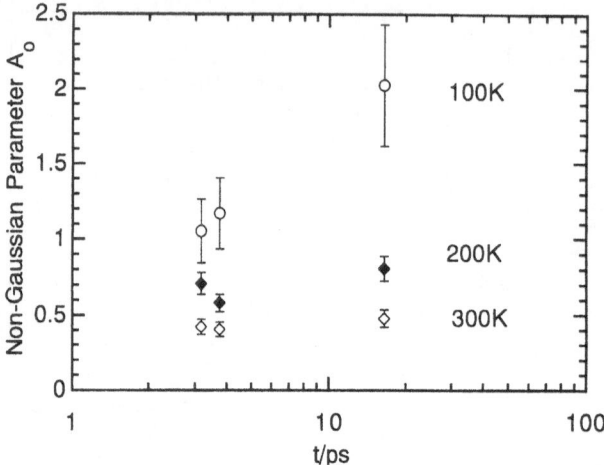

Fig. 32. Time-dependence of non-Gaussian parameter A_0 for PMMA at 100 ($\bigcirc$), 200 ($\blacklozenge$), and 300 K ($\diamondsuit$). Time is defined as $h/\delta\varepsilon$. (Reprinted with permission from [140]. Copyright 1998 Elsevier Science B. V., Amsterdam)

pose measurements of the elastic scattering intensity of PMMA (T_g=378 K) were performed with three different energy resolutions $\delta\varepsilon$ of 0.25, 1.1, and 1.3 meV [140]. Defining time as $h/\delta\varepsilon$ for convenience, the non-Gaussian parameters A_0 for PMMA are shown in Fig. 32 as a function of time for three temperatures below T_g. A_0 increases with time at 100 K while it is almost independent of time at 200 and 300 K. In the theoretical calculation [143, 144], increase of the non-Gaussian parameter is observed below T_g in the whole time region, and near T_g it is almost time independent. The experimental observations agree qualitatively with the theoretical prediction and suggest that the non-Gaussian parameter arises from heterogeneity of vibration motions. Above T_g, the theory also predicts that the non-Gaussian parameter A_0 decreases with time in the long time region. This decrease in A_0 has been observed in polybutadiene [142] above T_g in the time region of 0.05 to 0.5 ns.

6.5
Fragility Dependence of Non-Gaussian Parameter

The concept of fragility was introduced in order to characterize the temperature dependence of viscosity or relaxation time of the α-process τ_α of glass-forming materials above T_g [145, 146]. In Fig. 33, viscosities for many glass-forming materials are plotted against the inverse of temperature normalized to T_g. These temperature dependencies are generally described by the Vogel-Fulcher equation while the detail of the behavior depends on kinds of materials. Some glass-forming materials show the Arrhenius-like behavior while others show a large deviation. Such difference is characterized by a concept of fragility. According to Angell [145], glass-forming materials showing the Arrhenius-like behavior are

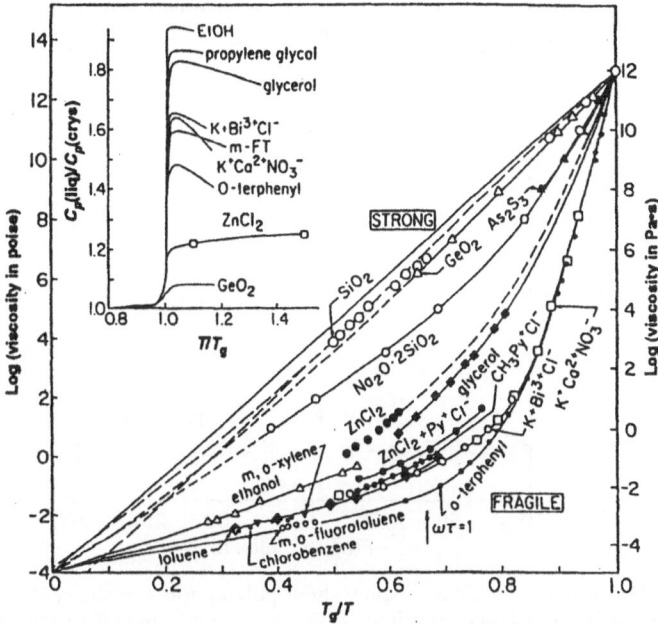

Fig. 33. Plot of viscosities of various glass-forming materials against T/T_g (Angell plot). *Inset* shows jump in heat capacity C_p at T_g, which is generally large for the *fragile* liquids and small for *strong* liquids. (Reprinted with permission from [146]. Copyright 1995 American Association for the Advancement of Science)

classified into *strong* and those showing a large deviation into *fragile*. Conveniently, it is quantified by a fragility index m which is defined as an apparent activation energy of viscosity or τ_α at T_g:

$$m = \left[\frac{d(\log \eta)}{d(T_g/T)} \right]_{T=Tg} = \left[\frac{d(\log \langle \tau_\alpha \rangle)}{d(T_g/T)} \right]_{T=Tg} \tag{43}$$

Fragility represents a property of the α-process in the supercooled state. On the other hand, the motion which is mainly reflected in the non-Gaussian parameter A_0 is a harmonic vibration, at least, below T_g because the mean square displacement $\langle u^2 \rangle$ is proportional to temperature T (see Fig. 30). The *fragility* and the non-Gaussian parameters are very different properties of amorphous polymers; the former is related to the α-process in the supercooled state and the latter is related to the vibrational motion below T_g. It was reported that some properties in glassy states such as heat capacity (see inset of Fig. 33) and boson peak intensity [147] could be related to the *fragility*. In a similar sense, the non-Gaussianity or the degree of heterogeneity must be related to the *fragility*.

The non-Gaussian parameter is shown in Fig. 34 as a function of fragility index m which was taken from [148] for reduced temperatures $T/T_g = 0.5$ and 0.8.

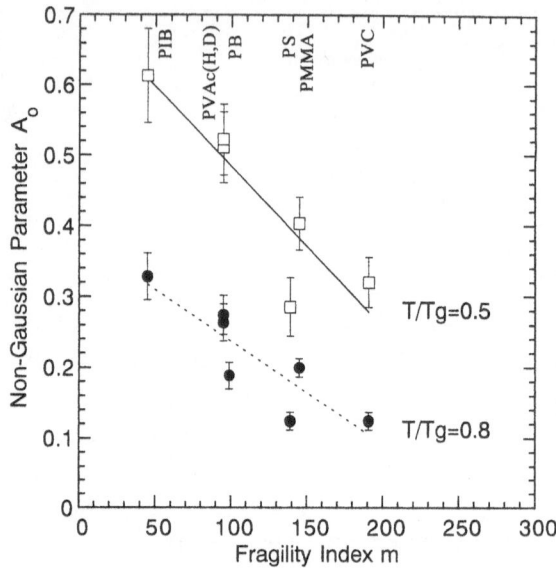

Fig. 34. Non-Gaussian parameter A_0 of amorphous polymers as a function of fragility index m at $T/T_g=0.5$ (□) and 0.8 (●). (Reprinted with permission from [140]. Copyright 1998 Elsevier Science B.V., Amsterdam)

Although the data points are rather scattered, there is no doubt that the non-Gaussian parameter A_0 decreases as the polymers become more fragile. In other words, the stronger (or less fragile) the polymer the more heterogeneous it is.

6.6
Origin of the Non-Gaussian Parameter

As mentioned in Sect. 6.1, the leading origin of the non-Gaussian behavior must be the dynamical heterogeneity due to local different environments for each molecule [53]. Finally we consider other factors affecting the non-Gaussianity. In these measurements, motions reflected in the non-Gaussian parameter are harmonic vibrations, at least, at temperatures below T_g. As shown in inelastic X-ray scattering experiments [58–60], there coexist the localized mode (the boson peak) and the extended mode (Debye mode) in the low energy region. This coexistence must cause the non-Gaussian behavior because of the distribution of mean square distributions. Furthermore, it should be noted that contribution of the microscopic anisotropy of the motions to the non-Gaussian behavior is not negligible because the non-Gaussian behavior was observed even in the crystalline phase of polyethylene [149]. As for the anharmonic effect, it may be neglected below T_g because the harmonic behaviors such as the Bose-scaling of the inelastic scattering and the T-linear dependence of the mean square displacement are sustained.

7
Concluding Remarks

We have reviewed recent inelastic and quasielastic neutron scattering studies on dynamics of glass-forming materials in the glassy and supercooled state, especially focusing on glass-forming polymers. These experimental works were stimulated by some microscopic theories such as the mode coupling theory (MCT) [7, 72], the trapping diffusion theory [143, 144] and the coupling theory [150], and computer simulations and vice versa. MCT especially plays an important role in the development of this field, which is based on the fundamental equation of the Mori-Zwanzig formula. From a technical point of view, recent rapid developments of inelastic and quasielastic neutron scattering techniques enabled us to do such investigations. Many current topics were identified in these studies; the boson peak in the low energy region which is an origin of the anomalous heat capacity in the low temperature region at around 10–20 K, the fast process in picosecond order, the E-process related to local conformational transitions of a polymer chain, and the dynamical heterogeneity of glass-forming polymers. These features of glass-forming materials, apart from the E-process, are very general and commonly observed in most glass-forming materials irrespective of kinds of materials, revealing the essential features of glass transition which is a universal phenomenon in all materials.

On the other hand, from traditional polymer science we know that glass transition is related to some typical natures of polymers such as chain flexibility, which is related to the glass transition temperature T_g. It has recently been revealed by Inoue and Osaki [151] that the chain flexibility is also related to the dynamical heterogeneity. The studies reviewed in this article do not take into account such polymer effects. Studies taking into account such effects are essential to connect the recent microscopic studies on glass transition with the traditional studies in polymer field. Hence study along this direction is one of the next steps in polymer field.

References

1. Doi M, Edwards SF (1986) The theory of polymer dynamics. Clarendon Press, Oxford
2. Richter D, Dianoux AJ, Petry W, Teixeira J (ed) (1989) Dynamics of disordered materials. Springer Proceedings in Physics 37. Springer, Berlin Heidelberg New York
3. Dianoux AJ, Petry W, Richter D (eds) (1993) Dynamics of disordered materials II, physica A210. North-Holland, Elsevier, Amsterdam, Grenoble
4. Ngai KL, Riande E, Wright GB (eds) (1993) The Second International Discussion Meeting on Relaxation in Complex Systems; J Non-Cryst Solid, 172/174 (1994) North-Holl and Elsevier, Amsterdam, Alicante
5. Odagaki T, Hiwatari Y, Matsui J (eds) (1997) Yukawa International Seminar 1996 (YKIS'96), Prog Theor Phys Suppl 126 (1997), Yukawa Institute for Theoretical Physics:Kyoto, Kyoto
6. Ngai KL, Riande E, Ingram MD (eds) (1997) The Third International Discussion Meeting on Relaxation in Complex Systems, J Non-Cryst Solid, 235/237 (1998), North-Holl and Elsevier:Amsterdam, Vigo

7. Götze W (1991) In: Hansen JP, Levesque D, Zinn-Justin J (eds) Liquid, freezing and glass transition. North-Holland, Amsterdam
8. Turchin VF (1965) Slow neutrons (translated from Russian), Israel Program for Scientific Translations, Jerusalem
9. Boutin H, Yip S (1968) Molecular spectroscopy with neutrons. The M.I.T. Press, Massachusetts
10. Willis BTM (ed) (1973) Chemical applications of thermal scattering. Oxford University Press, London
11. Springer T (1972) Quasielastic neutron scattering for the investigation of diffusive motions in solids and liquids. Springer, Berlin Heidelberg New York
12. Lovessey SW (1984) Theory of thermal neutron scattering from condensed matter. Clarendon Press, Oxford
13. Bee M (1988) Quasielastic neutron scattering. principles and applications in solid state chemistry, biology and materials science. Adam Hilger, Bristol
14. Higgins JS, Benoit HC (1994) Polymers and neutron scattering. Clarendon Press, Oxford
15. van Hove L (1954) Phy Rev 95:249
16. Bacon GE (1975) Neutron diffraction. Clarendon Press, Oxford
17. Rouse PE (1953) J Chem Phys 12:1272
18. Zimm BH (1965) J Chem Phys 24:269
19. DeGennes PG (1967) Physics 3:37
20. DeGennes PG, Dubois-Violette E (1967) Physics 3:181
21. Phillips WA (ed) (1981) Amorphous solids – low temperature properties. Springer, Berlin Heidelberg New York
22. Zeller RC, Pohl RO (1971) Phys Rev B6:2029
23. Phillips WA (1972) J Low Temp Phys 7:351
24. Anderson PW, Halperin BI, Varma CM (1972) Philos Mag 25:1
25. Buchenau U, Nücker N, Dianoux AJ (1984) Phys Rev Lett 53:2316
26. Buchenau U, Prager M, Nücker N, Dianoux AJ, Ahmad N, Phillips WA (1986) Phys Rev B34:5665
27. Buchenau U (1989) In: Richter D, Dianoux AJ, Petry W, Teixeira J (eds) Dynamics of disordered materials. Springer Proceedings in Physics. Springer, Berlin Heidelberg New York
28. Buchenau U, Galperin YM, Gurevich VL, Schober HR (1991) Phys Rev B43:5039
29. Kanaya T, Kaji K, Ikeda S, Inoue K (1988) Chem Phys Lett 150:334
30. Inoue K, Kanaya T, Ikeda S, Kaji K, Shibata K, Misawa M, Kiyanagi Y (1991) J Chem Phys 95:5332
31. Kanaya T, Kaji K, Inoue K (1992) Physica B 180/181:814
32. Kanaya T, Kawaguchi T, Kaji K (1993) J Chem Phys 98:8262
33. Frick B, Richter D (1993) Phys Rev B47:14,795
34. Kanaya T, Kawaguchi T, Kaji K (1994) J Non-Cryst Solids 172/174:327
35. Kanaya T, Kawaguchi T, Kaji K (1996) J Chem Phys 105:4342
36. Frick B, Buchenau U, Richter D (1995) J Coll Polym Sci 273:421
37. Yamamuro O, Matsuo T, Takeda K, Matsuo T, Kanaya T, Kawaguchi T, Kaji K (1996) J Chem Phys 105:732
38. Yamamuro O, Tsukushi I, Matsuo T, Takeda K, Kanaya T, Kaji K (1997) J Chem Phys 106:2997
39. Hansen J, Kanaya T, Nishida K, Kaji K, Tanaka K, Yamaguchi A (1998) J Chem Phys 108:6492
40. Viras F, King TA (1984) Polymer 25:899
41. Viras F, King TA (1984) Polymer 25:1411
42. Malinovsky VK, Sokolov AP (1986) Solid State Comm 57:757
43. Malinovsky VK, Novikov VN, Sokolov AP (1988) Chem Phys Lett 143:111
44. Gochiyaev VZ, Malinovsky VK, Novikov VN, Sokolov AP (1991) Phylo Mag B 63:777

45. Krüger M, Soltwisch M, Petscherizin I, Quitmann D (1992) J Chem Phys 96:7352
46. Duval E, Boukenter A, Achibat T (1990) J Phys Condens Matter 2:10,227
47. Achibat T, Boukenter A, Duval E, Lorentz G, Etienne S (1991) J Chem Phys 95:2949
48. Tucker JE, Reese W (1967) J Chem Phys 46:1388
49. Kanaya T, Imai M, Kaji K (1996) Physica B226:82
50. Ramos MA, Vieira S, Bermejo FJ, Dawidowski L, Fischer HE, Shhober H, Gonzalez MA, Loong CL, Price DL (1997) Phys Rev Lett 78:82
51. Kanaya T, Zorn R, Tsukushi I, Murakami S, Kaji K, Richter D (1998) J Chem Phys 109:10,456
52. Kanaya T, Miyakawa M, Kawaguchi T, Kaji K (1995) In: Giordano M, Leporini D, Tosi MP (eds) Proceedings of Workshop on Non-Equilibrium Phenomena in Supercooled Fluids, Glasses and Amorphous Materials. World Scientific, Singapore
53. Buchenau U, Pecharroman C, Zorn R, Frick B (1996) Phys Rev Lett 77:659
54. Laird BB, Schober HR (1991) Phys Rev Lett 66:636
55. Schober HR, Laird BB (1991) Phys Rev B44:6746
56. Schober HR, Oligschleger C (1996) Phys Rev B53:11,469
57. Tsukushi I, Kanaya T, Kaji K (1998) J Non-Cryst Solids 235/237:250
58. Masciovecchio C, Ruocco G, Sette F, Krisch M, Verbeni R, Bergmann U, Soltwisch M (1996) Phys Rev Lett 76:3356
59. Mermet A, Cunsolo C, Vuval D, Krisch M, Masciovecchio C, Perghem S, Ruocco G, Sette F, Verbeni R, Viliani G (1998) Phys Rev Lett 80:4205
60. Sette F, Krisch M, Masciovecchio C, Ruocco G, Monaco G (1998) Science 280:1550
61. Horbach J, Kob W, Binder K (1998) J Non-Cryst Solids 235/237:320
62. Nakayama T (1998) Phys Rev Lett 80:1244
63. Derrida B, Orbach R, Yu K-W (1984) Phys Rev 29:6645
64. Rosenberg H (1985) Phys Rev Lett 54:704
65. Dianoux AJ, Page JN, Rosenberg HM (1987) Phys Rev Lett 58:886
66. Yakubo K, Courtens E, Nakayama T (1990) Phys Rev B42:1078
67. Mermet A, Surovtsev NV, Duval E, Jal JF, Dupuy-Philon J, Dianoux AJ (1996) Europhys Lett 36:277
68. Ramos MA, Buchenau U (1998) In: Esquinazi P (ed) Tunneling systems in amorphous and crystalline solids. Springer, Berlin Heidelberg New York
69. Gil L, Ramos MA, Bringer A, Buchenau U (1993) Phys Rev Lett 70:182
70. Nakayama T (1999) J Phys Soc Japan 68:40
71. Ediger MD, Angell CA, Nagel SR (1996) J Phys Chem 100:13,200
72. Götze W, Sjögren L (1992) Prog Rept. Phys 55:241
73. Mezei F, Knaak W, Farago B (1987) Phys Rev Lett 58:571
74. Knaak W, Mezei F, Farago B (1988) Europhys Lett 7:529
75. Fujara F, Petry W (1987) Europhys Lett 4:921
76. Bartsch E, Fujara F, Kiebel M, Sillescu H, Petry W (1989) Ber Bunsenges Phys Chem 93:1252
77. Petry W, Bartsch E, Fujara F, Kiebel M, Sillescu H, Farago B (1991) Z Phys B – Condensed Matter 83:175
78. Fischer EW, Meier G, Rabenau T, Patkowski A, Steffen W, Thönnes W (1991) J Non-Cryst Solids 131/133:134
79. Frick B, Richter D, Petry W, Buchenau U (1988) Z Phys B – Condensed Matter 70:73
80. Richter D, Frick B, Farago B (1988) Phys Rev Lett 61:2465
81. Frick B, Farago B, Richter D (1990) Phys Rev Lett 64:2921
82. Richter D, Zorn R, Farago B, Frick B, Fetters LJ (1992) Phys Rev Lett 68:71
83. Richter D, Arbe A, Colmenero J, Monkenbusch M, Farago B, Faust R (1998) Macromolecules 31:1133
84. Roessler E, Warschewske U, Eiermann P, Sokolov AP, Quitmann D (1994) J Non-Cryst Solids 172/174:113
85. Kaji K, Urakawa H, Kitamaru R, Inoue K, Kiyanagi Y (1982) KENS Report 3:38

86. Kaji K, Urakawa H, Kitamaru R, Inoue K, Kiyanagi Y (1982) KENS Report 3:40
87. Inoue K, Ishikawa Y, Watanabe N, Kaji K, Kiyanagi Y, Iwasa H, Kohgi M (1985) Nucl Instr Meth. A 238:401
88. Kanaya T, Kawaguchi T, Kaji K (1992) Physica B 182:403
89. Frick B, Richter D, Trevino S (1993) Physica A201:88
90. Frick B, Richter D, Zorn R, Fetters LJ (1994) J Non-Cryst Solids 172/174:272
91. Buchenau U, Schönfeld C, Richter D, Kanaya T, Kaji K, Wehrmann R (1994) Phys Rev Lett 73:2344
92. Zorn R, Arbe A, Colmenero J, Frick B, Richter D, Buchenau U (1995) Phys Rev E52:781
93. Kanaya T, Ishida T, Kawaguchi T, Kaji K (1995) Physica B213/214:502
94. Kanaya T, Kawaguchi T, Kaji K (1996) J Chem Phys 104:3841
95. Kanaya T, Kaji K, Bartos J, Klimova M (1997) Macromolecules 30:1107
96. Kiebel M, Bartsch E, Debus O, Fujara F, Petry W, Sillescu H (1992) Phys Rev B45:10,301
97. Wuttke J, Kiebel M, Bartsch J, Fujara F, Petry W, Sillescu H (1993) Z Phys B91:357
98. Wuttke J, Hernandez J, Li G, Goddens G, Cummins HZ, Fujara F, Petry W, Sillescu H (1994) Phys Rev Lett 72:3052
99. Roessler E, Sokolov AP, Eiermann P, Warschewske U (1993) Physica A 201:237
100. Berry GC, Fox TG (1968) Adv Polym. Sci 5:261
101. Ferry J (1980) Viscoelastic properties of polymers. Wiley, New York
102. Donth E (1982) J Non-Cryst Solids 53:325
103. Johari GP, Goldstein M (1970) J Chem Phys 53:2372
104. Johari GP (1973) J Chem Phys 58:1766
105. Recent MD simulation by Odagaki has shown that rotational motions of anisotropic molecules are origins of the JG process
106. Ediger MD (1991) Annu Rev Phys Chem 42:225
107. Gapinski J, Steffen W, Patkowski A, Sokolov AP, Kisliuk A, Buchenau U, Russina M, Mezei F, Schober H (1999) J Chem Phys 110:2312
108. Floudas G, Higgins JS, Kremer F, Fischer EW (1992) Macromolecules 24:4955
109. Steffen W, Patkowski A, Meier G, Fischer EW (1992) J Chem Phys 96:4171
110. Yano O, Wada Y (1971) J Polym Sci Part A 9:669
111. Mermet A, Duval E, Surovtsev NV, Jal JF, Dianoux AJ, Yee AF (1997) Europhys Lett 38:515
112. Colmenero J, Arbe A (1998) Phys Rev B57:113,508
113. Wigner EP (1958) Ann Math 67:325
114. Cohen MH, Turnbull D (1959) J Chem Phys 31:1164
115. Buchenau U, Zorn R (1992) Europhys Lett 18:523
116. Kanaya T, Tsukushi I, Kaji K, Bartos J, Kristiak J (1999) Phys Rev E 60:1906
117. Schatzki TF (1962) Polym Sci 57:337
118. Valeur B, Jarry J-P, Geny F, Monnerie L (1975) J Polym Sci Polym Phys Ed 13:667
119. Valeur B, Monnerie L, Jarry J-P (1975) J Polym Sci Polym Phys Ed 13:675
120. Helfand E, Wasserman ZR, Weber TA (1979) J Chem Phys 70:2016
121. Helfand E, Wasserman ZR, Weber TA (1980) Macromolecules 13:526
122. Takeuchi H, Roe RJ (1991) J Chem Phys 94:7446
123. Takeuchi H, Roe RJ (1991) J Chem Phys 94:7458
124. Boyd RH, Gee RH, Han J, Jin Y (1994) J Chem Phys 101:788
125. Gee RH, Boyd RH (1994) J Chem Phys 101:8028
126. Kim E-G, Mattice WL (1994) J Chem Phys 101:6242
127. Moe NE, Ediger MD (1996) Polymer 37:1787
128. Moe NE, Ediger MD (1996) Macromolecules 29:5484
129. Kanaya T, Kaji K, Inoue K (1991) Macromolecules 24:1826
130. Kanaya T, Kawaguchi T, Kaji K (1999) Macromolecules 32:1672
131. Alvarez F, Alegria A, Colmenero J (1991) Phys Rev B44:7306
132. Alvarez F, Alegria A, Colmenero J (1993) Phys Rev B47:125
133. Singwi KS, Sjolander A (1960) Phys Rev 119:863

134. Donth E (1991) J Non-Cryst Solids 131/133:204
135. Fischer EW, Donth E, Steffen W (1992) Phys Rev Lett 68:2344
136. Matsuoka S, Quan X (1991) Macromolecules 24:2770
137. Arbe A, Colmenero J, Monkenbusch M, Richter D (1998) Phys Rev Lett 81:590
138. Sillescu H (1999) J Non-Cryst Solids 243:81
139. Kanaya T, Tsukushi I, Kaji K (1997) Prog Theor Phys Supply 126:133
140. Kanaya T, Tsukushi I, Kaji K, Gabrys B, Bennington SM (1998) J Non-Cryst Solids 235/237:212
141. Rahman A, Singwi KS, Sjölander A (1962) Phys Rev 126:986
142. Zorn R (1997) Phys Rev B 55:6249
143. Odagaki T, Hiwatari Y (1990) Phys Rev A41:929
144. Odagaki T, Hiwatari Y (1991) Phys Rev A43:1103
145. Angell CA (1988) J Phys Chem Solids 49:863
146. Angell CA (1995) Science 267:1924
147. Sokolov AP, Rössler E, Kisliuk A, Quitmann D (1993) Phys Rev Lett 71:2062
148. Böhmer R, Ngai KL, Angell CA, Plazek DJ (1993) J Chem Phys 99:4201
149. Kanaya T, Buchenau U, Koizumi S, Tsukushi I, Kaji K (2000) Phys Rev B 61: R 6451
150. Ngai KL (1979) Comment Solid State Phys 9:121
151. Inoue T, Osaki K (1996) Macromolecules 29:1595
152. Nagano K, Kanaya T, Fukunaga T, Mizutani U (1996) Funtai Yakin 43:726

Editor: Prof. T. Kobayashi
Received: December 1999

The Mesoscopic Theory of the Slow Relaxation of Linear Macromolecules

Vladimir N. Pokrovskii

Department of Physics, University of Malta, Msida, MSD 06, Malta
e-mail: vpok@isaac.phys.um.edu.mt

The review is devoted to the description of the relaxation behaviour of very concentrated solutions or melts of linear polymers. A mesoscopic approach, which deals with the dynamics of a single macromolecule among others and is based on some statements of a general kind, is used. From a strictly phenomenological point of view, the mesoscopic approach is a microscopic macromolecular approach. It reveals the internal connection between phenomena and gives more details than the phenomenological approach. From a strictly microscopic point of view, it is a phenomenological one. It needs some mesoscopic parameters to be introduced and determined empirically. However, the mesoscopic approach permits us to explain the different phenomena of the dynamic behaviour of polymer melts – diffusion, neutron scattering, viscoelasticity, birefringence and others – from a macromolecular point of view and without any specific hypotheses. The mesoscopic approach constitutes a phenomenological frame within which the results of investigations of behaviour of weakly-coupled macromolecules can be considered. The resultant picture of the thermal motion of a macromolecule in the system appears to be consistent with the common ideas about the localisation of a macromolecule: one can introduce an intermediate length which has the sense of a tube diameter and/or the length of a macromolecule between adjacent entanglements. In fact, it appears to be the most important parameter of the theory, as it was envisaged by Edwards and by de Gennes.

Keywords. Birefringence, Localisation of macromolecule, Polymer dynamics, Viscoelasticity, Weakly-coupled chains

Advances in Polymer Science, Vol. 154
© Springer-Verlag Berlin Heidelberg 2001

1
Introduction

Polymers differ from other substances by the size of their molecules which, appropriately enough, are referred to as macromolecules, since they consist of thousands or tens of thousands of atoms (molecular weight up to 10^6 or more) and have a macroscopic rectilinear length (up to 10^{-4} cm). The atoms of a macromolecule are firmly held together by valence bonds, forming a single entity. In polymeric substances, the weaker van der Waals forces have an effect on the components of the macromolecules which form the system. The structure of polymeric systems is more complicated than that of low-molecular solids or liquids, but there are some common properties: the atoms within a given macromolecule are ordered, but the centres of mass of the individual macromolecules and parts of them are distributed randomly. Remarkably, the mechanical response of polymeric systems combines the elasticity of a solid with the fluidity of a liquid. Indeed, their behaviour is described as viscoelastic, which is closely connected with slow (relaxation time to 1 s or more) relaxation processes in systems. Relaxation phenomena in very concentrated solutions and melts of polymers have been a prime focus of research during recent decades. The theory of phenomena is based on the fundamental principles of statistical physics. However, the peculiarities of the structure and the behaviour of the system necessitate the implementation of special methods and heuristic models which are different from those for gases and solids. This makes the subject of general interest for physicists and chemists.

The general theory of the relaxation behaviour of polymer solutions and melts appears to be derived from the universal models of long macromolecules which can be applied to any flexible macromolecule notwithstanding the nature of its internal chemical structure. Although many universal models are useful in the explanation of the behaviour of a polymeric system, the theory which will be described in this work is based on the coarse-grained model of a flexible macromolecule, the so-called, bead-and-spring or subchain model [1,2]. In the foundation of this model, one finds a simple idea to observe the dynamics of a set of representative points (beads, sites) along the macromolecule instead of observing the dynamics of all the atoms. It has been shown that each point can be considered as a Brownian particle, so the theory of Brownian motion can be applied to the motion of a macromolecule as a set of linear-connected beads. The large-scale or low-frequency properties of macro-molecules and macromolecular systems can be universally described by this model. The results do not depend on the arbitrary number of sites.

The work reviews the results of the theory of the relaxation behaviour of very concentrated solutions or melts of linear polymers, mainly the results of the mesoscopic approach, which deals with the dynamics of a single macro-molecule among others. The tradition of the mesoscopic approach begins with the first work on concentrated polymer solutions (for a short history of ap-proaching to the problem see [3]). The classical experience of studying a se-parate macromolecular coil in a viscous liquid by Rouse [2], Cerf [4], Zimm [5], Peterlin [6] and many others appears to be very important for formulating the theory of the behaviour of concentrated solutions and melts of polymers. Some specifying hypotheses about the behaviour of the probe macromolecule in the system were formulated. One of the hypothesis assumes that the motion of the chain is essentially confined in a tube-like region made of the surround-ing macromolecules [7]. The reptation motion of macromolecule inside the tube was introduced by de Gennes [8] and the application to viscoelasticity was elaborated by Doi and Edwards [9]. Another hypothesis, which was pro-posed by Edwards and Grant [10] ascribes to the environment of a probe macromolecule the properties of a relaxing medium. The subsequent theory, based on the non-Markovian stochastic equation, was elaborated by many re-searchers (for short review see [11]) and led to the formal mesoscopic theory of the relaxation of macromolecules. Though different approaches emerged as alternative approaches, they describe the same object – thermal motion of macromolecules in a system of weakly coupled macromolecules – and, in fact, support each other. The localisation of a macromolecule in a tube, postulated by Edwards and by de Gennes, follows from the formal mesoscopic approach. On the other hand, the reptation motion has to be taken into account to ex-plain large-scale conformational relaxation and mobility of a macromolecule.

Section 1 of the work can be considered as an introduction to elementary polymer physics. I have felt it necessary to give a brief account of the basis of the subject to make the review self-contained. Basic terms and heuristic mod-els of polymer science are described in this Section. The subsequent sections contain an exposition of the mesoscopic theory itself, while in Sects. 5, 6 and 7 consequences of the theory when compared to experimental data on diffusion, neutron scattering, viscoelasticity and optical anisotropy are discussed.

2
Macromolecules in Equilibrium

2.1
Macromolecular Coil

One says that the microstate of a macromolecule is determined if a sequence of atoms, the distances between atoms, valence angles, the potentials of interactions and so on are determined. The statistical theory of long chains developed in considerable detail in monographs [12, 13] defines the equilibrium quantities that characterise a macromolecule as a whole as functions of the macromolecular microparameters.

To say nothing about atoms, valence angles and so on, one can notice that the length of a macromolecule is much larger than its breadth, so one can consider the macromolecule as a flexible, uniform, elastic thread with coefficient of elasticity a, which reflects the individual properties of the macromolecule [13, 14]. Thermal fluctuations of the macromolecule should give the dependence of the mean square end-to-end distance $\langle R^2 \rangle_0$ on the length of macromolecule M and temperature T, which is, we believe, measured in energy units. If $MT \gg a$

$$\langle R^2 \rangle_0 = \frac{2Ma}{T}. \tag{1}$$

The last relation shows that a long macromolecule rolls up into a coil at high temperatures. The smaller the elasticity coefficient a is, the more it coils up.

Another name for the model of flexible thread is the model of persistence length or the Kratky-Porod model. The quantity a/T is called the persistence length [12].

One can use another way to describe the long macromolecule. One can see that at high temperatures there is no correlation between the orientations of the different parts of the macromolecule which are not close to each other along the chain. This means that the chain of freely-jointed rigid segments reflects the behaviour of a real macromolecule. This model carries the name of Werner Kuhn who introduced it in his pioneering work [15].

The expression for the mean square end-to-end distance can be written as the mean square displacement of a Brownian particle after z steps of equal length l

$$\langle R^2 \rangle_0 = zl^2. \tag{2}$$

If we return to the chain, z is the number of Kuhn segments in the chain, and l is the length of the segment. These quantities are uncertain yet. To avoid uncertainty, one adds a condition which is usually

$$zl = M. \tag{3}$$

Equations (2) and (3) determine the model of a freely-jointed segment chain which is frequently used in polymer physics as a microscopic heuristic model [16].

In such a way, there are two universal, (that is, irrespective of the chemical nature) methods of description of a macromolecule; either as a flexible thread or as freely-jointed segments. Either model reflects the properties of each macromolecule long enough to be flexible. A relation

$$\frac{2a}{T} = l$$

follows from the comparison of Eqs. (1) – (3). This relation demonstrates the imperfection of either model when applied to a real macromolecule. Indeed, it shows that the length of a segment or the elasticity coefficient depends on the temperature, which contradicts the proposed features of the models.

In any case, the mean square end-to-end distance of a long macromolecule $\langle R^2 \rangle_0$ is small compared to the length of the macromolecule. Whatever its chemical composition, a macromolecule which is long enough rolls up into a coil as a result of thermal motion, so that its mean square end-to-end distance becomes proportional to its molecular length

$$\langle R^2 \rangle_0 \sim C_\infty(T)M. \tag{4}$$

The temperature dependence of the size of a macromolecular coil is included in the coefficient of stiffness $C_\infty(T)$ which has the sense of the ratio of the squared length of a Kuhn segment to the squared length of the chemical bond, and can be calculated from the local chemical architecture of the chain. The results of the calculations were summarised by Birstein and Ptitsyn [12] and by Flory [13].

The probability distribution function for the fixed end-to-end distance R of macromolecule can be written down on either ground. In the simplest case, it is the Gaussian distribution

$$W(R) = \left(\frac{3}{2\pi \langle R^2 \rangle_0} \right)^{\frac{3}{2}} \exp\left(-\frac{3R^2}{2\langle R^2 \rangle_0} \right). \tag{5}$$

We may note that function (5) has the unrealistic feature that R can be larger than the maximum extended length M of the chain. Though more realistic distribution functions are available [12, 13], in this paper, function (5) is sufficient for our purpose.

2.2
Bead-and-Spring Model

A macrostate of a macromolecule can always be described with the help of the end-to-end distance $\langle R^2 \rangle_0$. To give a more detailed description of the macromolecule, one should use a method introduced by the pioneering work reported by Kargin and Slonimskii [1] and by Rouse [2] whereby the macromolecule is divided into N subchains of length M/N. The points at which the subchains join to form the entire chain (the beads) will be labelled 0 to N respectively and their positions will be represented by $r^0, r^1, \ldots, r^N$. If one assumes that each subchain is also sufficiently long, and can be described in the same way as the entire chain, then the equilibrium probability distribution for the positions of all the particles in the macromolecule is determined by the multiplication of N distribution functions of the type of (5)

$$W(r^0, r^1, \ldots, r^N) = C \exp(-\mu A_{\alpha\gamma} r^\alpha r^\gamma), \tag{6}$$

where

$$\mu = \frac{3}{2b^2} = \frac{3N}{2\langle R^2 \rangle_0}, \tag{7}$$

and the matrix $A_{\alpha\gamma}$ has the form

$$A_{\alpha\gamma} = \begin{Vmatrix} 1 & -1 & 0 & \ldots & 0 \\ -1 & 2 & -1 & \ldots & 0 \\ \ldots & \ldots & \ldots & \ldots & \ldots \\ \ldots & \ldots & \ldots & \ldots & \ldots \\ 0 & 0 & 0 & \ldots & 1 \end{Vmatrix}. \tag{8}$$

One notes that the free energy of a macromolecule in this approach is given by

$$F(\mathbf{r}^0, \mathbf{r}^1, \ldots, \mathbf{r}^N) = \mu T A_{\alpha\gamma} \mathbf{r}^\alpha \mathbf{r}^\gamma, \tag{9}$$

and this determines the force on the particle in the first order in r

$$K_i^\nu = -\frac{\partial F}{\partial r_i^\nu} = -2\mu T A_{\nu\gamma} r_i^\gamma \tag{10}$$

where ν is the bead number.

When it is determined in this way, the model is called the Gaussian sub-chain model: it can be generalised in a number of ways. When additional rigidity is taken into account, we have to add the interaction between different particles, so that matrix (8) is replaced, for example, by a five-diagonal matrix. It is also possible to take into account the finite extension of subunits by including in (9) terms of higher order in r, and so on.

The Gaussian subchain model and its possible generalisations are universal models which can be applied to every macromolecule which is long enough, irrespective of its chemical composition. The model does not describe the local structure of the macromolecule in detail, but describes correctly the property on a large length-scale. The described model plays a fundamental role in the theory of the equilibrium and non-equilibrium properties of the polymer.

The distribution function (6) allows one to calculate, in the Gaussian sub-chains approach, the equilibrium characteristics of a macromolecule; for example, the mean square radius of gyration of a macromolecular coil

$$\langle S^2 \rangle = \frac{1}{1+N} \sum_{\alpha=0}^{N} \langle (\mathbf{r}^\alpha - \mathbf{q})^2 \rangle, \quad \mathbf{q} = \frac{1}{1+N} \sum_{\alpha=0}^{N} \mathbf{r}^\alpha. \tag{11}$$

Note that at large N the variable

$$s = \frac{\alpha}{N+1}, \quad 0 \le s \le 1$$

can be introduced, and the matrix A expressed by (8) can be written down as

$$A \approx -\frac{1}{N^2} \frac{\mathrm{d}^2}{\mathrm{d}s^2}.$$

This allows one to rewrite expressions, considered here and later, in other forms. In this work, however, we prefer to use the discrete label.

2.3
Normal Co-ordinates

The equilibrium and non-equilibrium characteristics of the macromolecular coil are calculated conveniently in terms of new co-ordinates, so-called normal co-ordinates defined by

$$r^\beta = Q_{\beta\alpha}\rho^\alpha \quad \rho^\alpha = Q_{\alpha\gamma}^{-1}r^\gamma, \tag{12}$$

such that the quadratic form in equations (6) and (9) assumes a diagonal form, so that

$$Q_{\lambda\mu}A_{\lambda\gamma}Q_{\gamma\beta} = \lambda_\mu \delta_{\mu\beta}. \tag{13}$$

It can readily be seen that the determinant of the matrix given by (8) is zero, so that one of the eigenvalues, say λ_0, is always zero. The normal co-ordinate corresponding to the zeroth eigenvalue

$$\rho^0 = Q_{0\gamma}^{-1}r^\gamma$$

is proportional to the position vector of the centre of the mass of a macromolecular coil q, given by (11).

The behaviour of a macromolecule is conveniently described in a co-ordinate frame with the origin at the centre of the mass of the system. Thus $\rho^0 = 0$ and there are only N normal co-ordinates, numbered from 1 to N.

The transformation matrix Q can be chosen in a variety of ways which allow us to put extra conditions on it. Usually, it is assumed to be orthogonal and normalised. In this case, it can be demonstrated (see, for example, [17]) that the components of the transformation matrix are defined as

$$Q_{\alpha\gamma} = \left(\frac{2 - \delta_{0\gamma}}{N+1}\right)^{\frac{1}{2}} \cos\frac{(2\alpha + 1)\gamma\pi}{2(N+1)}. \tag{14}$$

For large N and small values of α, the eigenvalues are then given by

$$\lambda_\alpha = \left(\frac{\pi\alpha}{N}\right)^2, \quad \alpha = 0, 1, 2, \ldots, \ll N. \tag{15}$$

In the case of an orthogonal transformation, the relationship between the normal co-ordinate corresponding to the zeroth eigenvalue and the position of the centre of mass of the chain is

$$\rho^0 = q\sqrt{1 + N}. \tag{16}$$

The distribution function (6), normalised to unity, then assumes the following form

$$W(\rho^1, \rho^2, \ldots, \rho^N) = \prod_{\gamma=1}^{N} \left(\frac{\mu\lambda_\gamma}{\pi}\right)^{\frac{3}{2}} \exp(-\mu\lambda_\gamma \rho^\gamma \rho^\gamma).$$

The probability distribution function allows us readily to calculate the equilibrium moments of normal co-ordinates

$$\langle \rho_i^\nu \rho_k^\nu \rangle = \int W \rho_i^\nu \rho_k^\nu \{d\rho\} = \frac{1}{2\mu\lambda_\nu} \delta_{ik},$$

$$\langle \rho_i^\nu \rho_k^\nu \rho_s^\nu \rho_j^\nu \rangle = \frac{1}{4(\mu\lambda_\nu)^2} (\delta_{ik}\delta_{sj} + \delta_{is}\delta_{kj} + \delta_{ij}\delta_{ks}).$$

In the case of an orthogonal transformation, the mean square radius of gyration of the macromolecule (11) is expressed in equilibrium moments

$$\langle S^2 \rangle = \frac{1}{1+N} \sum_{\alpha=1}^{N} \langle \rho_i^\alpha \rho_i^\alpha \rangle.$$

The above formulae allow one to estimate the mean square radius of gyration of the macromolecule

$$\langle S^2 \rangle \approx \frac{1}{6} \langle R^2 \rangle_0.$$

In a case of a general transformation, relations (15) and (16) are not valid and ought to be replaced by other relations. A non-orthonormal transformation matrix was used when the non-equilibrium properties of the macromolecule in a liquid are investigated [5].

2.4
Excluded-Volume Effects

One says that the above results are valid for a non-interacting chain with zero volume. However, the monomers in a real macromolecule interact with each another, and this ensures, above all, that parts of the molecule cannot occupy the volume already occupied by other parts; i.e. the probabilities of successive steps are no longer statistically independent, as was assumed in the derivation of the above probability distribution functions and mean end-to-end distance [18]. The lateral interaction energy U depends on the differences of the position vectors of all particles of the chain and, in the simplest case, can be written as a sum of pair interactions

$$U = \sum_{\gamma \neq \nu} u(|r^\gamma - r^\nu|)$$

The short-range interaction between different particles can be approximated by the delta function

$$u(s) = vT\delta(s)$$

where v has the dimension of volume and is called the excluded volume parameter. The second virial coefficient $B(T)$, is considered to be, in this case, proportional to the excluded volume parameter

$$B(T) \sim v.$$

For the considered subchain model, the equilibrium distribution function that includes the particle interaction potential can be taken in the form

$$W = C \exp\left(-\mu A_{\alpha\gamma} r^\alpha r^\gamma - \frac{1}{T} U\right) \tag{17}$$

where C is the normalisation constant. The quantity μ is expressed through the mean end-to-end distance of a subchain b^2, whereby the lateral interactions are taken into account, so that it differs from expression (7) but nevertheless can be written on the basis of scaling speculations as

$$\mu \sim b^{-2}.$$

The free energy of a macromolecule in this case is given by

$$F(r^0, r^1, \ldots, r^N) = \mu T A_{\alpha\gamma} r^\alpha r^\gamma + U(r^0, \ r^1, \ldots, r^N) \tag{18}$$

However, if we are not interested in observing the variables r^0, $r^1, \ldots, r^N$ at all, the independent of these parameters free energy can be defined. This quantity can be calculated, starting from expression (17), so that it depends on the parameters T, N, b, v, whereas the arbitrary quantity N cannot influence the free energy of the macromolecular coil. When dimensional considerations taking into account, one has to write

$$F(T, b, v) = Tg(v/b^3)$$

The quantities v and b depend on arbitrary parameter N in such a way, that the quantity v/b^3 does not change.

The mean end-to-end distance of the entire chain $\langle R^2 \rangle$ is determined by lateral interactions, measured by the excluded volume parameter v. When dimensional considerations are taken into account [19], the quantity $\langle R^2 \rangle$ can be written in the form

$$\langle R^2 \rangle = b^2 f\left(N, v/b^3\right).$$

Of course, the end-to-end distance of the entire macromolecule $\langle R^2 \rangle$ does not depend on the arbitrary number of subchains N at $N \to \infty$. This means that the relation between $\langle R^2 \rangle$ and a finite number of subchains should be written in a way keeping the form of the relation under repeating divisions of the macromolecule. We have seen that the quantity v/b^3 does not depend on the number of divisions, so that the mean square end-to-end distance of the macromolecule is written as a power function

$$\langle R^2 \rangle = N^{2\nu} b^2.$$

It is easy to see that this relation is valid for an arbitrary number of divisions. Thus, a general consideration leads to the power dependence of the end-to-end distance of the macromolecule on its length

$$\langle R^2 \rangle \sim M^{2\nu}. \tag{19}$$

We can conclude that the dimensions of a macromolecular coil exhibiting excluded-volume effect are larger than those of the ideal coil, so that $\nu \geq 1/2$. However, it is necessary to fulfil a number of special and sophisticated calculations to find a specific value of power 2ν in expression (19) [20]. The first estimates of the index [18, 21] were done by simple self-consistent methods. Then the mean end-to-end distance was calculated by a perturbation method, while the chain in a imaginable 4-dimensional space is considered to be non-perturbed. The deviation of dimensionality of the imaginable space from the real physical space ϵ, is believed to be the small parameter of expansion. The first-order term gives [22] the following value of index

$$2\nu = \frac{9}{8}.$$

The answer is known to many decimal places [20].

We can conclude that the dimensions of a macromolecular coil exhibiting excluded-volume effect are larger than those of the ideal coil.

A great deal of effort has been expended in attempts to find a distribution function for the end-to-end length of the macromolecule [23]. Oono et al. [24] have shown that in the simplest approximation, the distribution function for dimensionless quantity $R^2/\langle R^2 \rangle$ is close to Gaussian, so the above results allow one to write down an approximate expression for the elasticity coefficient, when the excluded-volume effect is taken into account, in the form

$$\mu \sim \left(\frac{N}{M} \right)^{2\nu}. \tag{20}$$

In fact, the index in (20) is slightly different from 2ν.

We have already noted that the excluded-volume effect depends on the temperature which allows us to introduce θ-temperature at which the second virial coefficient is equal to zero. At high temperatures the repulsion interactions between particles prevail; on the contrary, at low temperatures the attraction interactions prevail, so that there is a temperature at which the repulsion and attraction effects exactly compensate each other. It is convenient to consider the macromolecular coil at θ-temperature to be described by expressions for an ideal chain, those demonstrated in Sects. 2.1 – 2.3. However, the old and more recent investigations [25, 26] demonstrate that the last statement can only be a very convenient approximation. In fact, the concept of θ-temperature appears to be immensely more complex than the above picture [18, 27].

2.5
Macromolecules in Solvent

The case considered in the previous section is idealised: the macromolecule does not exist in isolation but in a certain environment, for example, in a solution which is dilute or concentrated in relation to the macromolecules. The important characteristic for the case is the number of macromolecules per unit of volume n which can be written down through the weight concentration of polymer in the system c and the molecular weight (or length) of the macromolecule M

$$n = 6.026 \times 10^{23} \frac{c}{M} \quad cm^{-3}. \tag{21}$$

Macromolecules in dilute solutions ($c \ll 1$) can be considered as not interacting with each other. The mean distance between the centres of adjacent macromolecular coils $d \approx n^{-1/3}$ is much larger than the mean dimensions of the coil

$$d^2 \gg 2\langle S^2 \rangle_0, \tag{22}$$

where $\langle S^2 \rangle_0$ is the mean squared radius of gyration of the macromolecular coil.

Macromolecules in such solutions are considered usually not to aggregate, though this is not always valid [28]. Nevertheless, the behaviour of a single macromolecule has to be considered first of all. Now, the interaction of the atoms of the macromolecule with the atoms of solvent molecules has to be taken into account, apart from the interactions between the different parts of the macromolecule. To calculate the distribution function for the chain co-ordinate, one ought to consider $N + 1$ "big" particles of chain interacting with each other and each with "small" particles of solvent. One can anticipate that after eliminating the co-ordinates of the small particles in the distribution function, the distribution function of the chain co-ordinates can be taken in the form (17). In this case, the energy potential of the particle U is an effective potential, while taking into account both the interaction of the atoms of the macromolecule with the atoms of solvent and the interaction of the atoms of the macromolecule with each other. The results written down in the previous section are valid for the case considered.

From the energy point of view polymer–polymer contacts as compared with polymer–solvent contacts are preferred for some solvents called "good" solvents in this situation. A macromolecular coil swells and enlarges its dimension in a "good" solvent. On the contrary in a "bad" solvent, a macromolecular coil decreases its dimension and can collapse, turning into a condensed globule [18, 27].

The second virial coefficient of the macromolecular coil $B(T)$ depends not only on temperature but on the nature of the solvent. If one can find a solvent such that $B(T) = 0$ at a given temperature, then the solvent is called the θ-solvent. In such solvents, roughly speaking, the dimensions of the macromolecular coil are equal to those of an ideal macromolecular coil, that is the coil without particle interactions, so that relations of Sects. 2.1 – 2.3 can be applied in this case. However, it is a simplified description of the phenomenon. The fuller review of the theory of equilibrium properties of polymer solutions can be found in the monographs by des Cloizeaux and Jannink [29] and by Grossberg and Khokhlov [27].

2.6
Weakly-Coupled Macromolecules

Let us turn to the situation when the solution cannot be considered as dilute. Macromolecules in the solution entangle with each other and atoms of different macromolecules interact with each other through weak van der Waals forces.

The system of entangled macromolecules can exist in different physical states, depending on temperature. Further on, we will suppose that the temperature of the system exceeds the characteristic crystallisation and glass points, so that the system can be considered to be fluid. This is the case of concentrated polymer solutions or polymer melts. Any macromolecule in the system can only move as freely as its macromolecular neighbours allow it to.

The position of each macromolecule can be defined, as before, by specifying certain points along the macromolecule, spaced at distances that are equal, but not too small; as before, we shall refer to these points as particles. If one takes $N + 1$ points to define the position of the macromolecule, $3n(N + 1)$ co-ordinates are needed to specify the state of the entire system. The state of the system is described by the distribution of all the particles, and the corresponding equilibrium function is

$$W = C \exp\left(-\mu \sum_a A_{\gamma\nu} r^{a\gamma} r^{a\nu} - \frac{U}{T} \right), \tag{23}$$

where $r^{a\gamma}$ is the co-ordinate of the γth particle of a macromolecule labelled a or, in short, the co-ordinate of the particle $a\gamma$; μ and the matrix A are given by (7) and (8), respectively. The potential energy associated with the "'lateral'' interaction between the chains depends on the differences between the co-ordinates of all the particles in the system.

The distribution function for a single macromolecule and the mean dimensions of a macromolecular coil in the system are of particular interest. In contrast to the case considered in Sect. 2.4, the mean square end-to-end distance of the macromolecule is now a function of three dimensionless parameters

$$\langle R^2 \rangle = b^2 f(N, v/b^3, nb^3) \tag{24}$$

where b is the mean separation between neighbouring particles in a chain and $B(T)$ is the second virial coefficient.

In a dilute solution, the parameter nb^3 in (23) has a very slight effect. However, as the concentration of the polymer increases, the separation d between the coil centres decreases, and when

$$d = 2\langle S^2 \rangle^{\frac{1}{2}}$$

where $\langle S^2 \rangle$ is the mean square radius of gyration, the coils begin to overlap. This condition defines the critical molecular weight for a given concentration, or the critical concentration of the solution for a given molecular weight, at which overlap between domains occupied by macromolecules begins.

Furthermore, the increase in the concentration of the polymer is accompanied by the mutual interpenetrating of the coils. For concentrations approaching the limiting value $(c \to 1)$ the system of macromolecular coils becomes homogeneous in space; the macromolecular coils become entangled, and parts of different macromolecules are found at each point. This leads to a remarkable phenomenon – the screening of the lateral interactions between particles of a chosen chain – which was guessed by Flory [18] and strictly confirmed by Edwards in the mid seventies [9, p. 151]. Interactions between particles of a chosen macromolecule in a highly entangled system could be neglected; it means that, for description of the macromolecule, one can use the distribution function for ideal chain

$$W = C \exp(-\mu A_{\gamma\nu} r^{a\gamma} r^{a\nu}), \quad a = 1, 2, \ldots$$

which could be found by integrating (23) with respect to the co-ordinates of the particles of all the macromolecules apart from the particles of a chosen one. A fulfilment of such a procedure did really solve the problem.

The mean dimensions of the macromolecular coils in the entangled system are found to approach their unperturbed values, i.e. values they would have in a θ-solvent. The coil dimensions in the concentrated system are the same as the dimensions of ideal coils. This is confirmed by direct measurements of the dimensions of macromolecular coils in concentrated solutions and melts by neutron scattering [30, 31].

The entanglement of macromolecular coils in a concentrated system leads to a specific topological interaction between the macromolecules in the system, to the formation of sites and tangles [32, 33]. This interaction is particularly conspicuous in non-equilibrium phenomena.

Discussions of dynamic phenomena in polymer melts are frequently based on assumptions about the structure of the system, which is often taken to be a network with a characteristic site lifetime and nearest-neighbour separation [34]. A modification of this is the theory that postulates a certain intermediate scale, such as the diameter of a tube in which macromolecular displacement, i.e. reptation, is possible, but this hypothetical intermediate scale has been detected only in dynamic phenomena, and its existence should be regarded as a consequence rather than the origin of the theory. The theory that we shall consider in subsequent Sections does not rely on the assumption of an intermediate scale, but it does assume that the mean size $\langle R^2 \rangle$ and the macromolecular number density n (or concentration c) are the most significant parameters of the system. However, an intermediate dynamical length will appear in our theory later (see Sect. 5).

3
Dynamics of a Tagged Macromolecule

3.1
A Macromolecule in a Dilute Solution

3.1.1
Equation of Macromolecular Dynamics

As a preliminary step, we consider the dynamics of the macromolecular coil moving in the flow of a viscous liquid. The bead-spring model of a macromolecule is usually used to investigate large-scale or low-frequency dynamics of a macromolecular coil, while molecules of solvent are considered to constitute a continuum – viscous liquid. This is a mesoscopic approach to the dynamics of dilute solutions of polymers. There is no intention to collect all the available results and methods concerning the dynamics of a macromolecule in viscous liquid in this section. They can be found elsewhere [9,29]. We need to consider the results for dilute solutions mainly as a background to the discussion of the dynamics of a macromolecule in very concentrated solutions and melts of polymers.

Each bead of the chain is likened to a spherical Brownian particle, so that a set of the equation for motion for the macromolecule can be written as a set of coupled stochastic equations for coupled Brownian particles

$$m\frac{d^2 r^\alpha}{dt^2} = F^\alpha + G^\alpha + K^\alpha + \phi^\alpha, \quad \alpha = 0, 1, \ldots, N, \tag{25}$$

where m is the mass of a Brownian particle associated with a piece of the macromolecule of length $M/(N+1)$, r^α are the co-ordinates of the Brownian particles. Every Brownian particle is involved in thermal motion, which, as usual [9,35,36] can be described by putting a stochastic force ϕ^α (for a particle labelled α) into an equation of motion of a macromolecule. The essential features of the stochastic force are connected with properties of the dissipative forces F^α and G^α (the fluctuation-dissipation theorem). For the linear case the relation will be discussed at the end of Sect. 3.1.2.

According to relation (10) and (18), the elastic forces acting on the particle are taken in the form

$$K^\alpha = -\frac{\partial F}{\partial r^\alpha} = -2T\mu A_{\alpha\gamma} r^\gamma - \frac{\partial U}{\partial r^\alpha}, \tag{26}$$

whereas the dissipative forces F^α and G^α are needed in special discussion.

Each particle, moving at a velocity u^γ in the flow with constant velocity gradients $\nu_{ij} = \partial v_i/\partial x_j$, is acted upon by the hydrodynamic drag force, which is, generally speaking, determined by the relative motion of all the particles of the chain

$$F_j^\alpha = -\zeta B_{jl}^{\alpha\gamma}(u_l^\gamma - \nu_{li} r_i^\gamma) \tag{27}$$

where ζ is the coefficient of resistance of the particle in a viscous liquid. The matrix of hydrodynamic resistance $B_{jl}^{\alpha\gamma}$ depends on vectors of the relative distances between the particles $r^\gamma - r^\beta$.

On the deformation of the macromolecule, i.e. when the particles constituting the chain are involved in relative motion, an additional dissipation of energy takes place and intra-molecular friction forces appear [37]. The internal viscosity of the macromolecule is a consequence of the intramolecular relaxation processes occurring on the deformation of the macromolecule at a finite rate. The very introduction of the internal viscosity is possible only insofar as the deformation times are large, compared with the relaxation times of the intramolecular processes. If the deformation frequencies are of the same order of magnitude as the reciprocal of the relaxation time, these relaxation processes must be taken into account when considering the dynamics of the macromolecule and the dynamics of a dilute solution of the polymer. One must assume that the force acting on each particle is determined by the difference between the velocities of all the particles $u^\gamma - u^\beta$. These quantities must be introduced in such a way that dissipative forces do not appear on the rotation of the macromolecular coil as a whole. Thus the internal friction force must be formulated as follows, in terms of a general linear approximation with respect to velocities

$$G_i^\alpha = -\sum_{\beta\neq\alpha} C^{\alpha\beta}(u_j^\alpha - u_j^\beta)e_j^{\alpha\beta}e_i^{\alpha\beta}, \tag{28}$$

where $e_j^{\alpha\beta} = (r_j^\alpha - r_j^\beta)/|r^\alpha - r^\beta|$. Matrix $C^{\alpha\beta}$ is symmetrical, the components of the matrix are non-negative and may depend on the distance between the particles. The diagonal components of the matrix are equal to zero.

The internal friction force can conveniently be written in the form

$$G_i^\alpha = -G_{ij}^{\alpha\gamma} u_j^\gamma, \quad G_{ij}^{\alpha\beta} = \delta_{\alpha\beta}\sum_{\gamma\neq\alpha} C^{\alpha\gamma} e_i^{\alpha\gamma} e_j^{\alpha\gamma} - C^{\alpha\beta} e_i^{\alpha\beta} e_j^{\alpha\beta}. \tag{29}$$

The matrix $G_{ij}^{\alpha\gamma}$ is symmetrical with respect to the upper and lower indices and, in contrast to matrix $C^{\alpha\beta}$, has non-zero diagonal components. Expression (29) defines the general form of a matrix of internal friction, which allows the force to remain unchanged on the rotation of the macromolecular coil as a whole. Various internal-friction mechanisms, discussed in a number of studies [37–44] are possible. Their consideration leads to the determination of matrices $C^{\alpha\beta}$ and $G^{\alpha\beta}$ and to the nature of the dependence of the internal friction coefficients on the chain length and on the parameters of the macromolecule.

Comparison with experimental data demonstrates that the bead-spring model allows one correctly to describe the relaxation behaviour of dilute polymer solutions over a wide range of frequencies if the effects of excluded volume, hydrodynamic interaction, and internal viscosity are taken into account [45].

3.1.2
Linear Modes

Equations (25) – (29) determine the simplest approach to the dynamics of a macromolecule, even so, it appears to be rather complex if the effects of excluded volume, hydrodynamic interaction, and internal viscosity are taken into account. Due to these effects, all the beads in the chain ought to be considered to interact with each other in a non-linear way. To tackle with the problem, this set of coupled non-linear equations is usually simplified. There exist the different simpler approaches originating in works of Kirkwood and Riseman [46], Rouse [2], Zimm [5], Cerf [4], Peterlin [6] to the dynamics of a bead-spring chain in the flow of viscous liquid. The linearization is usually achieved by using preliminary-averaged forms of the matrix of hydrodynamic resistance (hydrodynamic interaction) [5] and the matrix of the internal viscosity [4]. In the last case, to ensure the proper covariance properties when the coil is rotated as a whole, Eq. (29) must be modified and written thus

$$G_j^\alpha = -G^{\alpha\gamma}(u_i^\gamma - \omega_{il}r_l^\gamma), \quad \omega_{ij} = \frac{1}{2}(\nu_{ij} - \nu_{ji}) \tag{30}$$

where $G^{\alpha\gamma}$ is now a symmetrical numerical matrix which is introduced phenomenologically.

In the simplest case the macromolecular coil can be considered to be perfectly drained and without internal friction [2]. These assumptions gives the simplest linear form of the dynamic equation

$$m\frac{d^2 r_i^\alpha}{dt^2} = -\zeta(\dot{r}_i^\alpha - \nu_{ij}r_j^\alpha) - 2T\mu A_{\alpha\gamma}r_i^\gamma + \phi_i^\alpha(t), \quad \alpha = 0,1,\ldots,N, \tag{31}$$

where r^α and $\dot{r}^\alpha$ are the co-ordinates and velocity of the Brownian particle, ζ is the friction coefficient of the particle in viscous liquid and $2T\mu$ is the coefficient of elasticity of the spring between adjacent particles. The matrix $A_{\alpha\gamma}$ depicts the connection of Brownian particles into the entire chain. When the transformation of the co-ordinates (12) is applied, the system of Eqs. (31) turns into a set of uncoupled equations for Rouse modes

$$m\frac{\mathrm{d}^2\rho_i^\alpha}{\mathrm{d}t^2} = -\zeta(\rho_i^\alpha - \nu_{ij}\rho_j^\alpha) - 2T\mu\lambda_\alpha\rho_i^\alpha + \xi_i^\alpha(t), \quad \alpha = 0, 1, \ldots, N, \tag{32}$$

where the eigenvalues of the matrix $A_{\alpha\gamma}$ for large N and small α are given by Eq. (15).

Dynamics (31) – (32) are commonly referred to as Rouse dynamics. It can be directly seen that the results for large-scale or low-frequency dynamics of a macromolecule do not depend on the number of particles N. In the over-damped regime ($m \to 0$), the mean sizes of the macromolecule relax to equilibrium values, whereas relaxation times are the Rouse relaxation times

$$\tau_\alpha^{\mathrm{R}} = \frac{\tau^*}{\alpha^2}, \quad \tau^* = \frac{\zeta b^2 N^2}{6\pi^2 T} \sim M^2, \tag{33}$$

where M is the length or the molecular weight of the macromolecule.

The random force ξ_i^γ in the dynamic Eqs. (32) is determined by its average moments and is selected from the condition that the equilibrium moments of the co-ordinates and velocities are known beforehand [35]. This requirement determines the relation

$$\langle \xi_i^\alpha(t)\xi_j^\gamma(t')\rangle = 2T\zeta\delta_{\alpha\gamma}\delta_{ij}\delta(t - t'), \tag{34}$$

which is valid to within first-order terms in the velocity gradients. Here and henceforth the angular brackets indicate averaging with respect to the assembly of realisations of the random force.

3.2
A Macromolecule in an Entangled System

3.2.1
Many-Chain Problem

Every macromolecule can be universally presented as consisting of z freely jointed rigid segments (Sect. 2.1). So, the system of entangled macromolecules can be imagined as consisting of nz interacting segments, every z of them being connected in chain, and the basis heuristic model, kinetics of which has to be investigated, is a system of interacting rigid segments connected in chains. In other words, it is a system of interacting Kuhn-Kramers chains. It seems to be a rather complex problem and it has no solution yet, though the problem was discussed by many investigators [47 – 50]. When we consider the relatively slow motion of the system, each macromolecule of the system can be schematically described as consisting of $N + 1$ linearly-coupled Brownian particles, as before. Thus we shall be able to look at the system as a suspension of $n(N + 1)$ interacting Brownian particles suspended in a viscous or viscoelastic liquid, which is made up of interacting segments. This is an heuristic model easier to consider. It is remarkable that the density of the Brownian particles is much less than the density of segments, so that the particles make up a weakly interacting system. However, the properties of the liquid in which the Brownian particles are suspended must be guessed when one is passing from the system of interacting segments to the more coarse model.

One can see that a particle located at point $r^{a\alpha}$ (where a is the label of the macromolecule to which the Brownian particle belongs, and α is the label of the particle in the macromolecule) is dragged with mean velocity $v_i^{\alpha} = \nu_{ik}r_k^{a\alpha}$, where ν_{ik} is the tensor of velocity gradients. The dynamics of each macromolecule can be described by equation which has the form of Eq. (25). We ought to add the forces of interaction with particles of the other macromolecules to the right-hand side of Eq. (25), so that the collective motion of the entire set of macromolecules (= set of Brownian particles) is described by a set of stochastic Markov equations which, for very slow motion, can be written in the form

$$m\frac{d^2 r_i^{a\alpha}}{dt^2} = -B_{a\alpha.b\beta}(\dot{r}_i^{b\beta} - \nu_{ij}r_j^{b\beta}) - 2\mu TA_{\alpha\gamma}r_i^{a\gamma} - \frac{\partial U}{\partial r_i^{a\alpha}} + \phi_i^{a\alpha}(t), \tag{35}$$

$$i = 1, 2, 3; \qquad \alpha = 0, 1, 2, \ldots, N; \qquad a = 1, 2, \ldots, n$$

where m is the mass of a Brownian particle associated with a section of the macromolecule of length M/N. The first term on the right is the hydrodynamic drag force, determined through the matrix $B_{a\alpha,b\beta}$, by all the Brownian particles in the system. It is assumed here that the particles are surrounded by a isotropic viscous not viscoelastic liquid which is valid for the cases when relaxation time of the segment liquid can be neglected. The second term represents the elastic force due to the nearest Brownian particles along the chain, and the third term is the direct interaction between all the Brownian particles. The last term represents the random thermal force whose statistical properties are, as usual, defined so that the equilibrium values of the calculated quantities are the same as those already known. The intramolecular friction forces (internal viscosity or kinetic stiffness) that arise when the macromolecular coil is deformed (see Sect. 3.1) are omitted because they are small in comparison to the other forces. The dynamics of interacting macromolecules is described by the system of Eqs. (35), and this was a starting point in a number of investigations of dynamics of weakly coupled macromolecules [51, 52].

A force acting on any part of the system gives rise to the agitation of the entire assembly of Brownian particles, so that, when the behaviour of the system and the mechanical forces are investigated, we have to consider the collective motion of all the particles in the same way that, for example, we examine the motion of the assembly of atoms in a solid [53]. Our task is therefore to find the normal co-ordinates of the considered polymer system, i.e. the variables that vary independently of one another. The mobility of the macromolecules is the main property of the system. When the system is agitated (mechanically or thermally), the macromolecules can readily exchange neighbours, but the integrity of each individual macromolecule remains unaffected. It allows to reduce the dynamics of the entire system to dynamics of single macromolecules. The validity of the mesoscopic approximation rests essentially on the fundamental experimental fact that quantities that characterise the behaviour of a polymer system have a well-defined single-valued dependence on the length of the macromolecule. The identification of the normal co-ordinates can be carried out in two stages, bearing in mind the particular properties of the system (the strong interaction along the chain and the weak interaction between macromolecules). The task of the first stage is to determine the dynamics of a single macromolecule, surrounded by all the others. Then, in the second stage, the normal co-ordinates of an individual macromolecule taken in its environment are determined.

To implement the first stage, we must eliminate all variables other than those that refer to the chosen macromolecule from the set of stochastic equations given by (35). We can imagine that there is a set of procedures which led to the equation of dynamics of a single macromolecule. A method of derivation of a single-chain equation is presented by Vilgis and Genz [52]. They used the Martin-Siggia-Rose formalism and demonstrated that several approximations are necessary to derive a single-chain equation for equilibrium

$$-\int_0^\infty \Gamma_{\alpha\gamma}(s)\dot{r}^\gamma(t-s)\mathrm{d}s - 2\mu T A_{\alpha\gamma} r^\gamma(t) = \phi^\alpha(t) \tag{36}$$

where the memory tensor function $\Gamma_{\alpha\gamma}(s)$ is connected with the correlation function of the random force $\phi^\alpha(t)$

$$\langle \phi^\alpha(t)\phi^\gamma(t')\rangle = 6T\Gamma_{\alpha\gamma}(t-t'). \tag{37}$$

An alternative way of deriving of the same relations was used by Schweizer [54], who, starting from the Louiville equation for the system, employed the Mori-Zwanzig projector operator techniques.

The value of these approaches to the problem is that they demonstrate the possibility of evaluating the memory function through the intermolecular correlation functions and structural dynamic factor of the system of interacting Brownian particles. Theoretical evaluations of the memory function are based on some approximations which, apparently, do not allow the calculation of quantities for the limiting case c^2M, but they are good in the cross-over region from Rouse to entanglement dynamics.

3.2.2
Linear Equation for a Probe Macromolecule

As far as the linear generalised Langevin equation of a single macromolecule is concerned, general speculations allow us to write down the form of the final results before the calculations are carried out. The requirements of covariance and of linearity in the co-ordinates and the velocities determine [55, 56] the general form of the equation for the dynamics of the tagged macromolecule:

$$
m\frac{d^2 r_i^\alpha}{dt^2} = -\int_0^\infty B_{ik}^{\alpha\gamma}(s)(\dot{r}_k^\gamma - \nu_{kj}r_j^\gamma)_{t-s}ds
$$
$$
-\int_0^\infty G_{ik}^{\alpha\gamma}(s)(\dot{r}_k^\gamma - \omega_{kj}r_j^\gamma)_{t-s}ds - 2\mu TA_{\alpha\gamma}r_i^\gamma + \phi_i^\alpha(t)
$$

(38)

where r^α is the co-ordinate of the particle of the tagged chain $(\alpha = 0, 1, \ldots, N)$.

The fundamental property of Eq. (38) is the time non-local form of the dissipative forces, represented by the first and second integral terms. The dependence of the memory functions on the top and bottom labels shows the influence of the other particles and the anisotropy of the medium on the motion of the considered particle. Strictly speaking, we should also write down these terms in the form of non-local expressions, since agitation directly through the chain propagates to a distance $\langle R^2 \rangle$, i.e. a distance that is large in comparison to the size of the Brownian particle under consideration. However, for the sake of simplicity, this will not be shown here, although the consequences of a non-local effect will be noted in Sect. 3.3.

The first term in the right-hand side of Eq. (38) has a form of the resistance force for a particle moving in viscoelastic liquid (Appendix A, Eq. (A.3)) which, in linear case, has to be considered to be isotropic. A remarkable property of the system under consideration is that the hydrodynamic interaction between the particles of the chain is negligible [51, 57, 58]. This allows one to introduce a scalar memory function

$$
B_{ik}^{\alpha\gamma}(s) = \beta(s)\delta_{\alpha\gamma}\delta_{ik}.
$$

The second dissipative term in (28) represents the intramolecular resistance due to change in the shape of the macromolecular coil. The force of intramolecular resistance appears when relative motion of the particles exists. It is determined by the non-linearity in the co-ordinate expression, which is similar to expression (28) for dilute solutions

$$
G_i^\alpha = -\int_0^\infty \sum_{\gamma \neq \alpha} C^{\alpha\gamma}(s)\left[e_i^{\alpha\gamma}e_j^{\alpha\gamma}(u_j^\alpha - u_j^\gamma)\right]_{t-s}ds.
$$

(39)

In contrast to the situation in dilute solutions, here the effect of intramolecular resistance is due to the entanglement of the macromolecules and to the motion of the particles in relation to one another. It is difficult to use expression (39) because of its non-linearity. To write down a linearised expression for the intramolecular resistance force, we refer to the speculations in Sect. 3.1. As before, we require the force to be independent of the rotation of the co-ordinate system with constant angular velocity. After simplifying (39), the resistance force can be similarly written as (30), in a form that is linear in the co-ordinates and velocities

$$G_i^\alpha = -\int_0^\infty G^{\alpha\gamma}(s)(u_i^\gamma - \omega_{ij}r_j^\gamma)_{t-s}\,ds, \quad \omega_{ij} = \frac{1}{2}(\nu_{ij} - \nu_{ji}). \tag{40}$$

The symmetrical matrix $G_{\alpha\gamma}$ represents the influence of particle γ on the movement of particle α. The only general requirement placed on matrix $G_{\alpha\gamma}(s)$ is that it must have a zero eigenvalue.

Thus, equation (38) takes the form

$$m\frac{d^2 r_i^\alpha}{dt^2} = -\int_0^\infty \beta(s)(\dot{r}_i^\alpha - \nu_{ij}r_j^\alpha)_{t-s}\,ds$$
$$-\int_0^\infty G^{\alpha\gamma}(s)(\dot{r}_i^\gamma - \omega_{ij}r_j^\gamma)_{t-s}\,ds - 2\mu T A_{\alpha\gamma}r_i^\gamma + \phi_i^\alpha(t) \tag{41}$$

This equation is the general equation for the dynamics of a single macromolecule in the case of linear dependence on the co-ordinates and velocities. However, the memory functions cannot be determined from general considerations: to calculate them one has to return to many-chain approaches. The value of these approaches to the problem is the possibility of evaluating the memory function through the intermolecular correlation functions and structural dynamic factor. Another way to solve the problem is to use some simple model considerations, as in works by Pokrovskii and Kokorin [59, 60].

As a reasonable assumption, it is possible to admit that each Brownian particle of the chain is in a similar situation and the intramolecular resistance force on a particle is determined equally by all the other particles of the chain, so that we can rewrite the memory function in (41) as

$$G^{\alpha\gamma}(s) = G_{\alpha\gamma}\,\varphi(s) \tag{42}$$

where $\varphi(s)$ is the scalar memory function and the matrix $G_{\alpha\gamma}$ can be approximated as

$$G_{\alpha\gamma} = \begin{Vmatrix} 1 & -1/N & \cdots & -1/N \\ -1/N & 1 & \cdots & -1/N \\ \cdots & \cdots & \cdots & \cdots \\ \cdots & \cdots & \cdots & \cdots \\ -1/N & -1/N & \cdots & 1 \end{Vmatrix}. \tag{43}$$

This matrix has a zero eigenvalue, when the transformation to normal coordinates (12) is fulfilled.

Let us note that Eq. (41) is a generalisation of the equation for the dynamics of a macromolecule in a dilute solution (Sect. 3.1): the effective viscous liquid in which the Brownian particle moves has been replaced by an effective viscoelastic liquid in the case of a concentrated solution; this introduces the concept of microviscoelasticity. Of course, if memory functions turn into δ-functions, Eq. (41) becomes identical to the equation of motion of the macromolecule in a viscous liquid.

3.3
Heuristic Speculations for Memory Functions

One has to start with the picture of the system of entangled macromolecules as a system of interacting segments and to guess that the relaxation of the mean orientation of the segments is a fundamental relaxation process. We can assume that in such a dense system of long macromolecules, the motion of a separate segment is determined strongly by its environment, being weakly dependent on its position in the chain. So, the mean orientation of segments $\langle e_i e_k \rangle$ can be introduced. We assume that a fundamental process has a relaxation time τ, so that the relaxation of orientation is described by equation

$$\frac{d\langle e_i e_k \rangle}{dt} = -\frac{1}{\tau}\left(\langle e_i e_k \rangle - \frac{1}{3}\delta_{ik}\right). \tag{44}$$

So, we assume the existence of a large single relaxation time

$$\tau \gg \tau^*$$

where τ^* is the largest Rouse relaxation time in "monomer" liquid. This assumption is consistent with the experimental evidence for very concentrated solutions of long macromolecules $(c^2 M \to \infty)$. So, one can write down for the memory functions in Eq. (41)

$$\beta(s) = 2\zeta\delta(s) + \frac{\zeta}{\tau}B\exp\left(-\frac{s}{\tau}\right), \quad \varphi(s) = \frac{\zeta}{\tau}E\exp\left(-\frac{s}{\tau}\right). \tag{45}$$

In formulae (44) and (45), ζ is the friction coefficient of a particle in a "monomer" liquid, while B and E are phenomenological parameters which will be discussed below. The correlation time τ can be interpreted as relaxation time of the mechanical (viscoelastic) reaction of neighbouring macromolecules, i.e. the system as a whole. This quantity will be calculated later (see Sect. 6), and the self-consistency of the theory will be demonstrated. Of course, such a choice of memory functions is eventually justified by empirical facts in later Sections, so we consider the memory functions (45) to be empirical, but to give a rather good description for the case $c^2 M \to \infty$.

To estimate the dependence of coefficient B in (45) on the length of the macromolecule, we consider, following Pokrovskii and Pyshnograi [61], very slow motion, for which the resistance-drag coefficient according to (A.5) can be written down as ζB. So, the dimensionless quantity B is a measure of the increase in the friction coefficient, due to the fact that the particle has behind it a wake of ambient macromolecules. Let us imagine the shear motion of the system as a motion of overlapping macromolecular coils each of which is characterised by the function of the mean number density of Brownian particles

$$\rho(r) = N\left(\frac{3}{2\pi\langle S^2\rangle_0}\right)^{3/2} \exp\left(-\frac{3r^2}{2\langle S^2\rangle_0}\right) \tag{46}$$

where r is the distance from the mass centre of the macromolecule.

The motion of a Brownian particle of the chosen macromolecule agitates a volume with a size of $\langle S^2\rangle_0^{1/2}$ through its adjacent chain particles. This volume is bigger the longer the macromolecules are. Note once more that the agitation comes through the chain, not through viscous friction. In this situation, for a particle with radius $a \ll \langle S^2\rangle_0^{1/2}$, the average environment has to be considered as a non-local liquid for which the following stress tensor can be written

$$\sigma_{ij} = -p\delta_{ij} + 2\int \eta(r-r')\gamma_{ik}(r')dr'. \tag{47}$$

The resistance-drag coefficient of a Brownian particle can be related to an influence function (see Appendix B) as

$$\zeta B = 6\pi a \int \eta(r)dr. \tag{48}$$

The influence function $\eta(s)$ can be found from simple considerations. When the system is deformed at velocity gradient γ_{ij}, two macromolecular coils, separated by a distance d_j, move beside each other at velocity $\gamma_{ij}d_j$. We add to the sum the contributions of every coil, apart from the chosen one, and find the density distribution of the energy dissipation for the chosen coil. The proportional coefficient depends only on the concentration of the Brownian particles if an assumption is made that local dissipation is determined by relative velocities

$$\eta(r)\gamma_{ij}\gamma_{ij} \sim \sum_a \rho(r)\rho(r-d^a)d_l^a d_k^a \gamma_{il}\gamma_{jk}.$$

When the linear approach is considered, the equilibrium density distribution function can be used. We turn the sum into the integral and, after calculating, obtain

$$\eta(r)\gamma_{ij}\gamma_{ij} \sim nN^2\left(\frac{3}{2\pi\langle S^2\rangle_0}\right)^{3/2}\left[(\langle S^2\rangle_0 - r^2)\delta_{lk} + r_l r_k\right] \times \exp\left(-\frac{3r^2}{2\langle S^2\rangle_0}\right)\gamma_{il}\gamma_{ik}. \tag{49}$$

So, the friction coefficient of the Brownian particle (48) can be estimated as

$$\zeta B \sim nN^2.$$

It means that $B \sim M^2$ and for sufficiently long chains $B \gg 1$.

The approximation of overlapping coils is very rough, nevertheless the index is not too different from the empirical value 2.4 (Sect. 6). This difference can be attributed to the fact that a macromolecule has to perform extra mo-

tions to disentangle itself from its neighbours. We can refer to old speculations by Bueche [62] and believe that by taking these speculations into account, the dependence of the proportional coefficient on the length of the
macromolecule can be improved. Thus we can write

$$B \sim M^\delta \tag{50}$$

and consider the index in (50) to have empirical value.

Note that the coefficients B and E are sensitive to the global architecture of
the macromolecule, that is whether the macromolecule is a ring, a brush, a
star, or something else. One can expect the index δ in (50) to be different in
each case.

3.4
Dynamics of a Reptating Macromolecule

So the first-order equation of the dynamics of a macromolecule in very concentrated solutions and melts of polymers has the form (41) where the memory functions defined by relations (42) and (45). This linear equation does not
include the reptation dynamics of a macromolecule introduced by de Gennes
[8] as a special type of anisotropic motion: a macromolecule moves along its
contour like a snake. Unbounded lateral motion is assumed to be completely
suppressed due to the entanglement of the tagged macromolecule with its
many neighbouring coils which, it is assumed, effectively constitutes a "tube"
of radius ξ. The reptation of a macromolecule is considered to be important to
describe the dynamics of solutions and melts of polymer [9].

To take the reptation of the macromolecule into account it is necessary to
introduce the anisotropy of the mobility for every particle (bead) in the considered coarse-grained model of a macromolecule consisting of N subchains.
We shall follow the work by Doi and Edwards [63] and Curtiss and Bird [47].
Although the original work by the latter authors deals with Kramers or rod-
bead models, we shall consider a simplified version which will allow us to
discuss the effect of reptation in a purified way.

To introduce the anisotropy of local mobility, we shall consider a part of the
chain, represented by a particle labelled α, to be characterised by a direction
vector with components

$$e_i^\alpha = \frac{r_i^{\alpha+1} - r_i^{\alpha-1}}{|r^{\alpha+1} - r^{\alpha-1}|}, \quad \alpha = 1, \ldots, N-1. \tag{51}$$

So there are $N - 1$ vectors for a chain, the zeroth and the last particles
having no direction.

Then, the tensor friction coefficient can be introduced for particles of the chain

$$\zeta_{ij}^{\alpha} = \zeta[(1+a)\delta_{ij} - ae_i^{\alpha}e_j^{\alpha}], \quad \alpha = 1, 2, \ldots, N-1, \tag{52}$$

$$\zeta_{ij}^{o} = \zeta_{ij}^{N} = \zeta\delta_{ij}.$$

One can see that at $a > 1$ the friction coefficient of a particle along the axis of a macromolecule is less than the friction coefficient in the perpendicular direction, so that the macromolecule can move more easily along its axis. At $a \to \infty$ the macromolecule does not move in the lateral directions at all.

Of course, it would be best to introduce the anisotropy of motion in equation (41). However, we follow Curtiss and Bird [47] and write the equation of macromolecular dynamics instead of (41) in the form

$$m\frac{d^2 r_i^{\alpha}}{dt^2} = -\zeta_{ij}^{\alpha}(\dot{r}_j^{\alpha} - \nu_{jl}r_i^{\alpha}) - 2\mu TA_{\alpha\gamma}r_i^{\gamma} + \phi_i^{\alpha}(t), \tag{53}$$

where ζ_{ij}^{α} is defined by Eq. (52). This is an equation of the form of Eq. (31) where the internal viscosity forces are omitted. No tube is used here. We can see that the anisotropy of the particle mobility makes the dynamic equation (53) non-linear, which fact makes it difficult for analysis. Statistical properties of the random force $\phi_i^{\alpha}(t)$ require a special discussion and approximation.

A very elegant linear model of reptating macromolecules was proposed by Doi and Edwards [63]. This model can be considered as a limiting case of the above model of Curtiss and Bird [47] at $a \to \infty$. It was proposed that, in this case of limiting anisotropy, the macromolecule moves inside the "tube" of radius ξ.

As before, we shall consider the chain of $N+1$ Brownian particles to be a schematisation of a macromolecule but, following Doi and Edwards [63], we assume that the distance between adjacent particles along the chain is equal to ξ, and that the number of particles is not arbitrary now, but satisfies the condition

$$N\xi^2 = \langle R^2 \rangle. \tag{54}$$

The states of the macromolecule will be considered in points of time in a time interval Δt, such that

$$\xi^2 = 6D_0\Delta t = \frac{6T}{N\zeta}\Delta t$$

where D_0 is the diffusion coefficient of the macromolecule in a monomeric viscous liquid.

Now, the stochastic motion of Brownian particles of the chain can be described by the equation for the particle co-ordinates

$$r^0(t + \Delta t) = \frac{1 + \phi(t)}{2} r^1(t) + \frac{1 - \phi(t)}{2} [r^0(t) + v(t)]$$

$$r^\nu(t + \Delta t) = \frac{1 + \phi(t)}{2} r^{\nu+1}(t) + \frac{1 - \phi(t)}{2} r^{\nu-1}(t), \quad \nu = 1, 2, \dots, N - 1$$

$$r^N(t + \Delta t) = \frac{1 + \phi(t)}{2} [r^N(t) + v(t)] + \frac{1 - \phi(t)}{2} r^{N-1}(t) \tag{55}$$

where $\phi(t)$ is a random quantity, which takes the values $+1$ or -1, and $v(t)$ is a vector of constant length ξ and random direction, so that

$$\langle \phi(t)\xi(u) \rangle = \delta_{tu} \quad \langle \phi(t) \rangle = 0$$
$$\langle v(t)v(u) \rangle = \delta_{tu}\xi^2, \quad \langle v(t) \rangle = 0. \tag{56}$$

The set of Eqs. (55) describes the stochastic motion of a chain. The "head" and the "tail" particles of the chain can choose random directions. Any other particle follows the neighbouring particles in front or behind.

We shall refer to the described models as the Curtiss-Bird model (Eq. (53)) and the Doi-Edwards model (Eq. (55)). There are different generalisations of these simple models, which were undertaken for the model to give a more accurate description of experimental evidence [64 – 70].

4
Low-Frequency Modes for Weakly-Coupled Macromolecules

4.1
Modes of Motion in Linear Approximation

Though the hypothetically "correct" equation for macromolecular dynamics must contain non-linear terms, it is useful to study the linear dynamics of a macromolecule based on Eq. (41). This equation ought to be regarded as the fundamental equation of the theory, and we can use normal co-ordinates (12) to get independent dynamic modes.

It is convenient to write down the zeroth normal co-ordinate, which describes the motion of the centre of mass of the macromolecular coil, separately, so the set of equations for the normal co-ordinates of the macromolecule now assumes the form

$$m \frac{d^2 \rho_i^0}{dt^2} = - \int_0^\infty \beta(s)(\dot{\rho}_i^0 - \nu_{ij}\rho_j^0)_{t-s} ds + \xi_i^0,$$

$$m \frac{d^2 \rho_i^\nu}{dt^2} = - \int_0^\infty \beta(s)(\dot{\rho}_i^\nu - \nu_{ij}\rho_j^\nu)_{t-s} ds - \int_0^\infty \varphi(s)(\dot{\rho}_i^\nu - \omega_{ij}\rho_j^0)_{t-s} ds$$

$$\qquad - 2\mu T \lambda_\nu \rho_i^\nu + \xi_i^\nu, \qquad \nu = 1, 2, \dots, N \tag{57}$$

where the eigenvalues λ_ν are defined by Eq. (15).

The formal representation of the solution of Eqs. (57) can be conveniently written in the form

$$\rho_i^\nu(t) = \int_0^\infty \left\{ \chi_\nu(s)\xi_i^\nu(t-s) + \left[\mu_\nu(s)\nu_{il}(t-s) + \pi_\nu(s)\omega_{il}(t-s)\right]\rho_l^\nu(t-s) \right\} ds,$$

(58)

where functions χ_α, μ_α and π_α are determined by their Fourier one-side transforms

$$\chi_\nu[\omega] = (2T\mu\lambda_\nu - m\omega^2 - i\omega B[\omega])^{-1},$$
$$\mu_\nu[\omega] = \beta[\omega]\chi_\nu[\omega], \quad \pi_\nu[\omega] = \varphi[\omega]\chi_\nu[\omega].$$

(59)

Functions χ_α, μ_α and π_α always vanish for $s \to 0$ and $s \to \infty$, if $m \neq 0$. Within the limits of applicability of the subchain model, the inertial effects have to be omitted, i.e. we can believe that $m = 0$, but this limit can change the values of functions (59) as functions of time s at limiting cases for $s \to 0$ and $s \to \infty$. To avoid any discrepancies, the results ought to be calculated at $m \neq 0$. Then the limiting values at $m \to 0$ can be obtained.

The expression for the velocity of the normal co-ordinate follows from Eq. (58). Differentiating (58) with respect to time, and integrating by parts, we find, by using the above shown properties of the integrands, that

$$\dot{\rho}_i^\nu(t) = \int_0^\infty \left\{ \dot{\chi}_\nu(s)\xi_i^\nu(t-s) + \left[\dot{\mu}_\nu(s)\nu_{il}(t-s) + \dot{\pi}_\nu(s)\omega_{il}(t-s)\right]\rho_l^\nu(t-s) \right\} ds.$$

(60)

Iteration of (58) and (60) can be used to expand the normal co-ordinates and their velocities into a power series of small velocity gradients of the medium. We can write down the zero-order approximation

$$\rho_{i0}^\nu(t) = \int_0^\infty \chi_\nu(s)\xi_i^\nu(t-s) ds$$

$$\dot{\rho}_{i0}^\nu(t) = \int_0^\infty \dot{\chi}_\nu(s)\xi_i^\nu(t-s) ds$$

(61)

and the first-order approximation

$$\rho_i^\nu(t) = \rho_{i0}^\nu(t) + \int_0^\infty \left[\mu_\nu(s)\nu_{il}(t-s) + \pi_\nu(s)\omega_{il}(t-s)\right]\rho_{l0}^\nu(t-s) ds$$

$$\dot{\rho}_i^\nu(t) = \dot{\rho}_{i0}^\nu(t) + \int_0^\infty \left[\dot{\mu}_\nu(s)\nu_{il}(t-s) + \dot{\pi}_\nu(s)\omega_{il}(t-s)\right]\rho_{l0}^\nu(t-s) ds.$$

(62)

Now, we can discuss in some detail the properties of the stochastic force $\xi_i^\alpha(t)$, defined so that $\langle \xi_i^\alpha(t) \rangle = 0$. The second-order moment

$$K_{ij}^{\alpha\gamma}(t,t') = \langle \xi_i^\alpha(t)\xi_j^\gamma(t') \rangle$$

(63)

depends on the velocity gradients and can be expanded into a power series of this quantity. The first-order term cannot, in general, satisfy the conditions of symmetry under interchange of the arguments of function (63), and must therefore be discarded. This means that, to within first-order terms in the velocity gradients, the correlation function has the same form as in the equilibrium, i.e. time-independent, case

$$K_{ij}^{\alpha\gamma}(t, t') = K_\alpha(t - t')\delta_{\alpha\gamma}\delta_{ij}.$$

The random force correlator is determined by the rule that, at equilibrium, the moments of the velocities and the co-ordinates must be known. In our simple case, the Fourier transform of the correlator is determined as follows

$$K(\omega) = \int_{-\infty}^{\infty} K(s)e^{i\omega s}ds = 2T\mathrm{Re}B[\omega], \tag{64}$$

where the one-sided Fourier transform of a function is indicated by square brackets

$$B[\omega] = \beta[\omega] + \varphi[\omega]. \tag{65}$$

4.2
Correlation Functions

All the observable physical quantities are average quantities which are expressed as some functions of moments of co-ordinates and velocities, first of all of second-order moments. These will be considered in this Subsection. To calculate these moments, we start with the expansion of normal co-ordinates and their velocities (62). For simplicity we shall omit the label of the normal co-ordinates in this section.

4.2.1
Equilibrium Correlation Functions

First, we shall consider moments at zero-velocity gradients; in other words, the equilibrium moment which depends on just one argument

$$\langle \rho_i(t)\rho_k(t - s)\rangle_0 = M(s)\delta_{ik}$$
$$\langle \dot\rho_i(t)\dot\rho_k(t - s)\rangle_0 = L(s)\delta_{ik}$$
$$\langle \rho_i(t)\dot\rho_k(t - s)\rangle_0 = S(s)\delta_{ik}.$$

The angle brackets denote the averaging over the ensemble of the realisation of the random forces in the particle equations of motion (57). We use relation (61) to write an expression for the moment of the normal co-ordinate

$$M(u) = \int_0^{\infty} \int_0^{\infty} \chi(s)\chi(v)K(u - s - v)ds\,dv, \tag{66}$$

where the correlation function of the random forces is defined by relation (64).

Multiplying (66) by $e^{i\omega u}$, and integrating with respect to u from $-\infty$ to ∞, we find the Fourier transform of the function which determines also the one-sided Fourier transform of this moment

$$M[\omega] = \frac{1}{2\mu\lambda} \frac{B[\omega] - im\omega}{2T\mu\lambda - m\omega^2 - i\omega B[\omega]}. \tag{67}$$

Note that the case when $m = 0$ can be considered, if $B(\omega) \neq 0$ at $\omega \to \infty$. Otherwise, to get the right results, it is essential to maintain the order in which the limit is approached, hence we ought to take $m = 0$ after calculation.

In a similar way, expressions for the one-sided Fourier transforms of the moment of velocities and the moment of co-ordinate and velocity can be found

$$L[\omega] = T\frac{i\omega}{2T\mu\lambda - m\omega^2 - i\omega B[\omega]}, \tag{68}$$

$$S[\omega] = \frac{T}{2T\mu\lambda - m\omega^2 - i\omega B[\omega]}. \tag{69}$$

4.2.2
One-point Non-Equilibrium Correlation Functions

The expressions for co-ordinates and velocities (62) with the same arguments can be used to make up proper combinations and, by averaging over the ensemble of realisation of random forces, we find the moments with accuracy to the first-order terms in velocity gradients

$$\langle \rho_i(t)\rho_k(t)\rangle = \frac{1}{2\mu\lambda}\delta_{ik} + 2\int_0^\infty \mu(s)M(s)\gamma_{ik}(t-s)ds$$

$$\langle \dot\rho_i(t)\dot\rho_k(t)\rangle = \frac{T}{m}\delta_{ik} + 2\int_0^\infty \dot\mu(s)\dot M(s)\gamma_{ik}(t-s)ds$$

$$\langle \rho_i(t)\dot\rho_k(t)\rangle = \int_0^\infty \Big\{\mu(s)S(s)\nu_{ik}(t-s) + \dot\mu(s)M(s)\nu_{ki}(t-s)$$

$$+ \pi(s)S(s)\omega_{ik}(t-s) + \dot\pi(s)M(s)\omega_{ki}(t-s)\Big\}ds. \tag{70}$$

We see that the non-equilibrium moments are expressed in terms of the equilibrium moments $M(s), L(s), S(s)$, which were determined in the previous Subsection by their Fourier transforms (67), (68) and (69). The last expression in (70) can be simplified when $m = 0$, because the inertial forces acting on the Brownian particles are unimportant. This is the only case which is of interest for application. In this case, the final expressions contain a function $R(x)$ of a non-negative argument x. The properties of the function are discussed in Appendix C.

By taking expressions (59) and (67) into account, we find the limiting relation

$$\pi(s) + \mu(s) + R(s) = 2\mu\lambda M(s). \tag{71}$$

As the first two moments in (70) are expressed in terms of the functions $M(s)$, and $\mu(s)$, it is convenient to express the third moment in these functions too. After calculating, we obtain

$$\langle\rho_i(t)\dot{\rho}_k(t)\rangle = \frac{1}{2\mu\lambda}\omega_{ki} + \int_0^\infty [\mu(s)\dot{M}(s) + \dot{\mu}(s)M(s)]\gamma_{ik}(t-s)\mathrm{d}s. \tag{72}$$

Thus, the non-equilibrium moments are determined by functions considered in the previous section.

4.2.3
Two-Point Non-Equilibrium Correlation Functions

Now, we turn to the calculation of two-point moments. We take the quantities defined by expression (62) and average the products of $\rho_i(t)$ and $\rho_k(t-s)$, $\rho_i(t)$ and $\dot{\rho}_k(t-s)$, respectively. By taking into account the properties of the equilibrium moments, we find that the final expressions for the moments take the form

$$\langle\rho_i(t)\rho_k(t-s)\rangle = M(s)\delta_{ik} \tag{73}$$
$$+ \int_0^\infty [\mu(u+s)M(u)+\mu(u)M(u+s)]\gamma_{ik}(t-s-u)\mathrm{d}u,$$

$$\langle\rho_i(t)\dot{\rho}_k(t-s)\rangle = S(s)\delta_{ik} + M(s)\omega_{ki}(t-s) \tag{74}$$
$$+ \int_0^\infty [\mu(u+s)\dot{M}(u)+\dot{\mu}(u)M(u+s)]\gamma_{ik}(t-s-u)\mathrm{d}u.$$

Naturally, the expressions determined in the previous section for one-point moments (70) and (72) follow, at $s = 0$, from formulae (73) and (74), respectively.

4.3
Relaxation Times of the Macromolecular Coil

The results discussed in the previous two sections do not depend on any concrete representations of the memory functions $\beta(s)$ and $\varphi(s)$. However, to calculate the correlation functions further, we must specify the functions $M_\nu(s)$ and $\mu_\nu(s)$ which are defined by their Fourier transforms (59) and (67). We obtain the results for the case, when the memory functions are given by their one-sided transforms

$$\beta[\omega] = \zeta\left(1 + \frac{B}{1 - i\omega\tau}\right), \qquad \varphi[\omega] = \frac{\zeta E}{1 - i\omega\tau}.$$

The correlation time τ and the dimensionless quantities B and E were discussed in Sect. 3.3, where the memory functions were introduced. So, according to formulae (59) and (67), when $m = 0$, one finds

$$\mu_\nu[\omega] = \frac{2(1 + B - i\omega\tau)\tau_\nu^R}{(1 - i\omega 2\tau_\nu^+)(1 - i\omega 2\tau_\nu^-)} \tag{75}$$

$$M_\nu[\omega] = \frac{1}{2\mu\lambda}\frac{2(1 + B + E - i\omega\tau)\tau_\nu^R}{(1 - i\omega 2\tau_\nu^+)(1 - i\omega 2\tau_\nu^-)}. \tag{76}$$

Here, τ_ν^R are the relaxation times of the macromolecule in a monomer viscous fluid – Rouse relaxation times

$$\tau_\nu^R = \frac{\tau^*}{\nu^2}, \qquad \tau^* = \frac{\zeta b^2 N^2}{6\pi^2 T}$$

and notations for the new sets of relaxation times are used

$$2\tau_\nu^\pm = \tau_\nu \pm \left(\tau_\nu^2 - 2\tau\tau_\nu^R\right)^{1/2}, \qquad \tau_\nu = \frac{\tau}{2} + \tau_\nu^R(1 + B + E). \tag{77}$$

Now, the reciprocal Laplace transform can be used to determine the functions $\mu(s)$ and $M(s)$ from (75) and (76). Before calculating, we remind the reader that the correct results can be obtained when the mass is retained in expressions (59) and (67). This changes expressions (75) and (76). However, it is easier to operate with limiting (at $m \to 0$) expressions. The final results can be improved by adding terms which contain function $R(t)$, described in Appendix C. We can also find, by simple alternative calculations, that

$$\mu_\nu(t) = T_\nu^+ \exp\left(-\frac{t}{2\tau_\nu^+}\right) - T_\nu^- \exp\left(-\frac{t}{2\tau_\nu^-}\right) - R(t) \tag{78}$$

$$M_\nu(t) = \frac{1}{2\mu\lambda_\nu}\left[S_\nu^+ \exp\left(-\frac{t}{2\tau_\nu^+}\right) - S_\nu^- \exp\left(-\frac{t}{2\tau_\nu^-}\right)\right] \tag{79}$$

where

$$T_\nu^\pm = \frac{\tau_\nu^R(1 + B) - \tau_\nu^\mp}{\tau_\nu^+ - \tau_\nu^-}, \qquad S_\nu^\pm = \frac{\tau_\nu^R(1 + B + E) - \tau_\nu^\mp}{\tau_\nu^+ - \tau_\nu^-}. \tag{80}$$

Expression (79) for the correlation function demonstrates that the relaxation times τ_ν^+ and τ_ν^- are the relaxation times of the mean square normal coordinates, or the mode labelled ν. There are two sets (branches) of relaxation times. One of the branches contains large relaxation times τ_ν^+, the other small. This is a characteristic feature of polymer melts, as is revealed in experiments. The complex dielectric permittivity of polar polymers is directly connected

[71, 72] with the correlation functions $M_\alpha(s)$, so measurements of the frequency dependence of dielectric permittivity reveal directly the relaxation times of the system. Two relaxation branches were discovered for cis-polyisoprene melts in experiments by Imanishi et al [73] and Fodor and Hill [74]. The fast relaxation times do not depend on the length of the macromolecule, while the slow relaxation times do. This is exactly the picture which qualitatively corresponds to the relaxation branches (77).

At large values of B the relation $\tau\tau_\nu^R \ll \tau_\nu^2$ is valid, so that the relaxation times (77) can be written in the form

$$\tau_\nu^+ = \tau_\nu - \frac{\tau\tau_\nu^R}{2\tau_\nu}, \qquad \tau_\nu^- = \frac{\tau\tau_\nu^R}{2\tau_\nu}. \tag{81}$$

In the limiting case of very large values of parameter B, i.e. $\zeta \to 0$ (but $\zeta B \neq 0$, $\zeta E \neq 0$), we find that

$$\tau_\nu^+ \to \tau_\nu, \qquad \tau_\nu^- \to 0.$$

In this limiting case, expressions (78) and (79) can be written as

$$\mu_\nu(t) = \frac{B\tau_\nu^R}{\tau_\nu} \exp\left(-\frac{t}{2\tau_\nu}\right) - \frac{B\tau_\nu^R}{\tau_\nu} R(t) \tag{82}$$

$$M_\nu(t) = \frac{1}{2\mu\lambda_\nu} \left[\frac{(B+E)\tau_\nu^R}{\tau_\nu} \exp\left(-\frac{t}{2\tau_\nu}\right) + \frac{\tau}{2\tau_\nu} R(t) \right].$$

Naturally, the latter formula is followed by the expression for the correlation function of the normal co-ordinates of the macromolecule in a viscous liquid (Rouse case)

$$M_\nu(t) = \frac{1}{2\mu\lambda_\nu} \exp\left(-\frac{t}{2\tau_\nu^R}\right). \tag{83}$$

5
The Localisation Effect

5.1
Mobility of a Macromolecule

We consider the diffusion of a macromolecule to be the diffusion of the co-ordinate of the centre of mass which is, according to relation (16), proportional to the zeroth normal co-ordinate, that is,

$$q \sim \rho^0.$$

We shall calculate the mean square displacement of the centre of mass of a diffusing macromolecule for a time t

$$\langle \Delta q^2 \rangle = \sum_{i=1}^{3} \langle [q_i(t) - q_i(0)]^2 \rangle.$$

Using the expression

$$q_i(t) - q_i(0) = \int_0^t \dot{q}_i(s)\,\mathrm{d}s$$

we obtain

$$\langle \Delta q^2 \rangle = \int_0^t \int_0^t \langle \dot{q}(s)\dot{q}(u) \rangle \mathrm{d}s\,\mathrm{d}u. \tag{84}$$

This reduces the calculations to the evaluation of the time-dependent velocity correlation function

$$L(u - s) = \frac{1}{3} \langle \dot{\rho}^0(s)\dot{\rho}^0(u) \rangle \sim \langle \dot{q}(s)\dot{q}(u) \rangle.$$

which appears to be specific for different cases (Sect. 4.2.1).

5.1.1
A Macromolecule in a Viscous Liquid

In this case, according to relations (68), the correlation function of the zero normal co-ordinate is determined by equation

$$L(x) = \frac{T}{m} \exp\left(-\frac{\zeta}{m}x\right)$$

which allows one, after simple calculations, to find that the macromolecule moves like a Brownian particle in a viscous liquid, and its displacement is given by the standard relation

$$\langle \Delta q^2 \rangle = 6D_0 t. \tag{85}$$

where the coefficient of the macromolecule diffusion is inversely proportional to the mobility of the macromolecular coil

$$D_0 \sim \frac{T}{\zeta_M}. \tag{86}$$

The dependence of the friction coefficient ζ_M of the macromolecule on its length M is connected with volume effects and effects of draining or non-draining (permittibility of macromolecular coils). It is known [9,19] that the description of the diffusion of macromolecular coils is in good agreement with experimental evidence for dilute solutions.

It has been shown [75, 76] that short macromolecules in melts at $M < M_e$, where M_e is 'the length of a macromolecule between adjacent entanglements' (see Sect. 6.4), can be considered to diffuse according to law (85) with the coefficient of diffusion

$$D_0 \sim M^{-1}.$$

5.1.2
A Macromolecule in an Entangled System

The mobility of a macromolecule constrained by other macromolecules can be also calculated as (84). The equation for the zeroth normal co-ordinate of the macromolecule can be taken as Eq. (57) at $\nu_{ij} = 0$. The one-sided Fourier transform velocity correlation function is determined by expression (71), so that we can write down the Fourier transform

$$L(\omega) = \frac{T}{-i\omega m + B[\omega]} + \frac{T}{i\omega m + B[-\omega]}.$$

Multiplying this expression by $\frac{1}{2\pi} e^{-i\omega t}$ and integrating with respect to ω from $-\infty$ to ∞, we find

$$L(t) = \frac{T}{\pi} \int_{-\infty}^{\infty} \frac{\cos \omega t}{-i\omega m + B[\omega]} \, d\omega.$$

We use the formula to write down the general expression for the mean square displacement of a particle in an arbitrary viscoelastic liquid

$$\langle \Delta q^2 \rangle = \frac{6T}{\pi(1+N)} \int_{-\infty}^{\infty} \frac{1 - \cos \omega t}{\omega^2(-i\omega m + B[\omega])} \, d\omega. \tag{87}$$

We shall now turn to a particular memory function and calculate the displacement in the simple case without inertia for which

$$B[\omega] = \zeta + \frac{\zeta B}{1 - i\omega\tau}.$$

We then find that the displacement of the centre of mass of a macromolecule is

$$\langle \Delta q^2 \rangle = \frac{6T\tau}{\zeta NB} \left(\frac{t}{\tau} + 1 - \exp\left(-\frac{t}{\tau}B\right) \right). \tag{88}$$

The displacement as a function of the ratio t/τ is shown in Fig 1. This function can be approximated as

$$\langle \Delta q^2 \rangle = \begin{cases} \dfrac{6T}{N\zeta} t, & t \ll \tau/B, \\[2mm] \dfrac{6T\tau}{N\zeta B}, & \tau/B \ll t \ll \tau, \\[2mm] \dfrac{6T}{N\zeta B} t, & t \gg \tau \end{cases} \tag{89}$$

$\lg \langle \Delta q^2 \rangle, \ \lg \Delta_{N/2}$

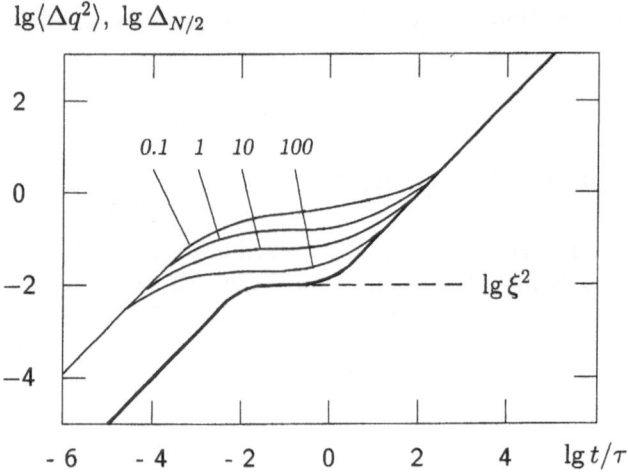

Fig. 1. Mobility of a macromolecule and its particles. The mean square displacement $\langle \Delta q^2 \rangle$ of the centre of mass of a chain (*thick solid line*) does not depend on parameter ψ, but the mean square displacement $\Delta_{N/2}$ of the central particle does. The values of parameter ψ are shown at the curves for $\Delta_{N/2}$. The theoretical curves are calculated according to formulae (89) and (95) for the values of the parameters: $B = 100$; $\chi = 10^{-2}$. The displacement is measured in units of $\left(\frac{\tau \langle R^2 \rangle}{\pi^2 \tau^*} \right)^{1/2}$. The picture demonstrates the existence of the intermediate scale ξ. Adapted from the papers of Kokorin and Pokrovskii [60, 77]

The mobility of the macromolecule changes sharply at $t = \tau/B$, but the displacement remains constant over a certain time of observation, and is given by

$$\xi^2 = \frac{6T\tau}{N\zeta B} = \frac{\langle R^2 \rangle \tau}{\pi^2 \tau^* B} = \frac{2}{\pi^2} \langle R^2 \rangle \chi. \tag{90}$$

The characteristic time τ/B and the characteristic scale thus appear in the theory. The dimensionless quantity χ can be interpreted as the ratio of the square of twice the characteristic scale to the mean square end-to-end distance of the macromolecule.

$$\chi = \frac{\tau}{2B\tau^*} = \frac{\pi^2}{8} \frac{(2\xi)^2}{\langle R^2 \rangle} \approx \frac{(2\xi)^2}{\langle R^2 \rangle}. \tag{91}$$

For short times of observation, $t \ll \tau/B$, the expression for the displacement is identical to (85) which was written for the displacement of a macromolecule in a viscous liquid. The situation is different for long observation times $t \gg \tau$. So, we can see that the coefficient of the diffusion of a macromolecule is different for different displacement: for distances which are less than ξ

$$D = \frac{T}{N\zeta}$$

and for distances which are bigger than ξ

$$D = \frac{T}{N\zeta B}. \tag{92}$$

In the last particular situation, the motion of a tagged chain is also coupled to the motion of neighbouring macromolecules, and the diffusion coefficient is determined both by the length of the test macromolecule M, and by the length of the ambient macromolecules M_0

$$D = D_0 B^{-1} \sim M_0^{-\delta} M^{-1}. \tag{93}$$

The self-diffusion coefficient for long chains is proportional to $M^{-\delta-1}$ and is small, and one then encounters a competing mobility mechanism that gives a different dependence of the diffusion coefficient on the length of the macromolecule. This will be discussed in Sect. 5.3.

5.2
Localisation of a Macromolecule

In the previous section the main differences between the thermal motion of a macromolecule in a viscous liquid and a macromolecule in a liquid which represents other macromolecules was discussed. It will be now shown, following Pokrovskii and Kokorin [60, 77, 78], that the mesoscopic linear equation of macromolecule dynamics is followed by a localisation effect.

Consider the mean square displacement of each particle of the chain

$$\Delta_\alpha(t) = \langle [r^\alpha(t) - r^\alpha(0)]^2 \rangle, \quad \alpha = 0, 1, \ldots, N. \tag{94}$$

It is convenient to transform the expression to normal co-ordinates and to separate the zeroth normal co-ordinate

$$\Delta_\alpha(t) = \frac{1}{N+1} \langle [\rho^0(t) - \rho^0(0)]^2 \rangle$$
$$+ 2 \sum_{\gamma=1}^{N} Q_{\alpha\gamma} Q_{\alpha\gamma} \left(\langle \rho^\gamma(t)\rho^\gamma(t) \rangle - \langle \rho^\gamma(t)\rho^\gamma(0) \rangle \right)$$

where the transformation matrix $Q_{\alpha\gamma}$ is defined by (14).

We use expression (79) to calculate the displacement and to find

$$\Delta_\alpha(t) = \langle \Delta \boldsymbol{q}^2 \rangle + 6 \sum_{\gamma=1}^{N} Q_{\alpha\gamma} Q_{\alpha\gamma} \frac{1}{2\mu\lambda_\gamma} \tag{95}$$

$$\times \left\{ S_\gamma^+ \left[1 - \exp\left(-\frac{t}{2\tau_\gamma^+} \right) \right] - S_\gamma^- \left[1 - \exp\left(-\frac{t}{2\tau_\gamma^-} \right) \right] \right\}.$$

The displacement of the centre of mass of the macromolecule $\langle \Delta \boldsymbol{q}^2 \rangle$ is defined by (89).

As an example, the time dependence of the displacement of the central particle is shown in Fig. 1 for certain values of the parameters. We can see that the dependence of any particle of the chain is similar to the dependence of the entire macromolecule. Both dependencies are characterised by the different mobility for short and long times of observation [77, 78].

For a short time of observation, $t \ll \frac{\tau}{B}$, the mobility of the particle is $N + 1$ times more than the mobility of the macromolecule

$$\Delta_\alpha(t) = \frac{6T}{\zeta} t. \tag{96}$$

In the internal interval $\frac{\tau}{B} < t < \tau$, when the change of the displacement is negligible, we can find the expression

$$\Delta_\alpha(t) = \frac{12\pi T \tau^*}{\zeta(N+1)} \left(\frac{B\chi}{B+E} \right)^{1/2}. \tag{97}$$

For a long time of observation, $t > \tau^* B$, the displacement of any particle is identical to the displacement of the entire macromolecule.

$$\Delta_\alpha(t) = \frac{6T}{\zeta(N+1)B} t. \tag{98}$$

Formula (97) defines a certain intermediate scale, which can be compared to the intermediate scale revealed in the consideration of the diffusion of a macromolecule (see expression (90)). We ought to believe that the local displacement of any point of the macromolecule should depend neither on the number of subchains nor on the length of the macromolecule. The displacement determines the single intermediate length defined by (90).

When we identify the quantities (90) and (97), we find the relation between the parameters of the theory at $B \gg 1$.

$$\psi = \pi^2 \frac{1}{\chi}. \tag{99}$$

The above value of ψ ensures that any Brownian particle of the chain does not move more than ξ during the times $t < \tau$. For this time of observation, the large-scale conformation of the macromolecule is frozen, but the small-scale motion of the particles confined to the scale ξ can take place, and the macromolecule, indeed, can be considered to be in a "tube" with radius ξ. The macromolecule wobbles around in the tube-like region, remaining near its initial position for some time (a time of localisation) which is the larger the longer the macromolecule is. A very long macromolecule appears, in fact, to behave exactly as if confined in a tube, though no other restrictions than equation (41) exist. Reptation of the macromolecule inside the tube is optional.

Equation (99) ought to be taken as an empirical relation, which is valid at $B \gg 1$. One can think that the more detailed theory is followed by the relation between the memory functions $\beta(s)$ and $\varphi(s)$. Relation (99) can be looked upon as a reflection of the connection between memory functions.

5.3
Mobility of a Macromolecule Through Reptation

Localisation of a macromolecule in a tube was assumed by Edwards [7] and by de Gennes [8]. The latter introduced reptation motion for the macromolecule to explain the law of diffusion of very long macromolecules in entangled systems. We consider the reptation of the macromolecule on the basis of the Doi-Edwards model described in Sect. 3.3.

As we saw in the previous section, a macromolecule in the system is confined by the scale ξ during the short time τ/B. To consider the reptation effect we can assume that the macromolecule moves only along its axis or inside a tube with a diameter 2ξ. The tube and its diameter are postulated in the earlier theories [9], but now we can believe that scale ξ defined by (90) is the radius of a tube.

To consider the mobility of the macromolecule, one ought to calculate the displacement of the centre of mass of the chain

$$q(t) = \frac{1}{N+1}\sum_{\nu=0}^{N} r^{\nu}(t). \tag{100}$$

After summing equations (55), we find

$$\Delta q(t) = q(t+\Delta t) - q(t) = \frac{1}{N}R(t)\phi(t) + \frac{1}{N}v(t)$$

We assume that $N \gg 1$ here, so $N+1 \approx N$, whereas $R = r^N - r^0$ is the end-to-end distance of the chain. Then, one calculates the correlation function

$$\langle \Delta q(t)\Delta q(u)\rangle = \frac{1}{N^2}\langle \phi(t)\phi(u)R(t)R(u)\rangle$$
$$+ \frac{1}{N^2}\langle v(t)v(u)\rangle$$
$$+ \frac{1}{N^2}[\langle \phi(t)R(t)v(u)\rangle + \langle \phi(u)R(u)v(t)\rangle].$$

Taking into account the properties (56) of random quantities, we find

$$\langle \Delta q(t) \Delta q(u) \rangle = \delta_{tu} \left(\frac{\xi^2}{N} + \frac{\xi^2}{N^2} \right).$$

Thus, the one-step mean square displacement is determined, at $N \gg 1$, by relation

$$\langle \Delta q^2 \rangle = \frac{\xi^2}{N}.$$

One multiplies the above quantity by the step number $\frac{t}{\Delta t} = \frac{6D_0}{\xi^2} t$, to find the mean square displacement in time t

$$\langle \Delta q^2 \rangle = \frac{\xi^2}{N} \frac{t}{\Delta t} = \frac{6D_0}{N} t. \tag{101}$$

Let us remind the reader that in the written relations the number of particles is not an arbitrary quantity but is defined by relation (54), so that the diffusion coefficient of the chain can be written as

$$D = D_0 \frac{\xi^2}{\langle R^2 \rangle_0}.$$

Remembering the definition of the internal scale (90), we write down the diffusion coefficient in another form

$$D = \frac{2}{\pi^2} D_0 \chi. \tag{102}$$

In the latter case the diffusion coefficient of the macromolecule does not depend on the length of the ambient macromolecules

$$D \sim M_0^0 M^{-2}. \tag{103}$$

Now, one can discuss the macromolecular-length dependence of the diffusion coefficient at $M > M_e$. We can see that the two mechanisms of the displacement of the centre of mass of the macromolecule are optional. The diffusion coefficients are defined by (93) and (103), respectively. The two competing mechanisms have a different length dependence of the self-diffusion coefficient

$$D \sim \begin{cases} M^{-\delta-1}, & non-reptation \\ M^{-2}, & reptation \end{cases}. \tag{104}$$

One can see from the comparison of equations (92) and (102) that the reptation motion of the macromolecules is revealed at the condition

$$\chi B \gg \frac{1}{2} \pi^2. \tag{105}$$

This relation determines a molecular weight M_c at which the mechanism of diffusion changes. Experimentally it has been shown that $M_c > M_e$, so there is a region where relation (93) for the diffusion coefficient is valid. The relation

(93) for this region (region of so-called constrained release) was obtained by Klein [79]. Of course, when the macromolecules are long enough, the reptation mechanism predominates, and this has indeed been confirmed experimentally [80, 81].

5.4
Quasi-Elastic Neutron Scattering

The investigation of macromolecule diffusion reveals that the time dependence of the mean square displacements of the centre of mass of the macromolecule and the chain particles are non-linear, so a dynamical internal length (a scale of localisation) can be introduced. Convincing confirmation of localisation effect is given by studying of neutron scattering. To observe the scattering on a single macromolecule in a system of entangled chains, the investigators have taken blends of chemically identical deuterated and non-deuterated macromolecules [82–84]. A small quantity of non-deuterated macromolecules among deuterated macromolecules determines scattering which can be considered to be scattering on single macromolecules.

An introduction to the theory of neutron scattering can be found, for instance, in a books by Hansen and Donald [85] and Higgins and Benoit [31]. The neutron scattering of a single macromolecule is determined by the scattering function

$$S(k, t) = \frac{1}{N+1} \sum_{\alpha, \gamma} \langle \exp[ik(r^\alpha(t) - r^\gamma(0))] \rangle. \tag{106}$$

The double sum is evaluated over all the Brownian particles of the macromolecule. In (106) k is the vector in the direction of the scattering, having the length

$$k = \frac{4\pi}{\lambda} \sin\frac{\theta}{2},$$

where λ is the wave-length of the initial beam and θ is the scattering angle. To investigate the motion of the internal parts of a macromolecular coil, the relation $k\langle R^2 \rangle^{1/2} \gg 1$, or $\lambda \ll \langle R^2 \rangle^{1/2}$ must be fulfilled. For typical macromolecules, it gives for the wave-length

$$\lambda \ll 10^{-6} cm.$$

To evaluate the scattering function (106), we can consider the expansion of the function

$$\langle \exp[ik(r^\alpha(t) - r^\gamma(0))] \rangle = 1 - \frac{1}{2} \sum_{i=1}^{3} k_i^2 \langle (r_i^\alpha(t) - r_i^\gamma(0))^2 \rangle - \ldots \tag{107}$$

and see that the function has real components only. Since the averaged values of the quantity $(r_i^\alpha(t) - r_i^\gamma(0))^2$ do not depend on the label of co-ordinate i, expression (106) can be written as

$$\langle \exp[ik(r^\alpha(t) - r^\gamma(0))] \rangle = \exp\left(-\frac{1}{6}k^2 \sum_{i=1}^{3} \langle (r_i^\alpha(t) - r_i^\gamma(0))^2 \rangle \right).$$

Hence, the scattering function (106) takes the form

$$S(k, t) = \frac{1}{N+1} \sum_{\alpha, \gamma} \exp\left(-\frac{1}{6}k^2 \sum_{i=1}^{3} \langle (r_i^\alpha(t) - r_i^\gamma(0))^2 \rangle \right). \tag{108}$$

To evaluate the non-coherent scattering function, we omit the correlation between particles with different labels, after which one has

$$S(k, t) = \frac{1}{N+1} \sum_{\alpha=0}^{N} \exp\left(-\frac{1}{6}k^2 \Delta_\alpha(t) \right) \tag{109}$$

where the mean square displacement $\Delta_\alpha(t) = \langle [r^\alpha(t) - r^\alpha(0)]^2 \rangle$ of the particle α is determined by expression (95).

We can easily see that all the properties of function $\Delta_\alpha(t)$ are reflected in the scattering function, which is shown in Fig. 2. In particular, a plateau is revealed on the plot of the scattering function. The value of the scattering

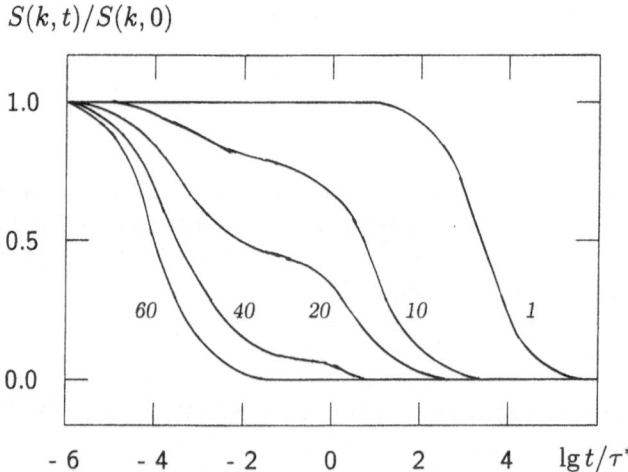

Fig. 2. Neutron scattering function. The function was calculated according to formula (109) at the values of the parameters: $B = 100$; $\chi = 10^{-2}$; $E = 10^3$, while values of $k\langle R^2 \rangle^{1/2}$ are shown at the curves. Adapted from the paper of Kokorin and Pokrovskii [77]

function in the plateau region is connected directly with the intermediate length ξ

$$S(k, t) = \exp\left(-\frac{1}{6}k^2\xi^2\right), \quad \frac{\tau}{B} < t < \tau. \tag{110}$$

Similar results are reported by Ronca [86] and des Cloizeaux [87]. Their calculations were based on different models which contain the tube diameter as an introduced parameter of the models.

Results on neutron scattering by specially prepared samples [83, 84] reveal the plateau region. The time dependence of the observed scattering functions appears to be very similar to the depicted scattering functions. The beam of neutrons characterised by k in the range from $0.058\,\text{Å}^{-1}$ till $0.135\,\text{Å}^{-1}$ was used by Richter et al [84]. It gives the values of $k\langle R^2\rangle^{1/2}$ from 5.8 to 13 for the typical size of macromolecular coils. We can see that the actual values expose the special time dependence revealing the intermediate length.

For an interpretation of the results and the calculation of the intermediate length, Richter et al [84] have used the scattering function calculated by Ronca [86]. Though we consider his calculations not to be quite consistent, the final results provide an estimate of the intermediate length – the tube diameter which is related to the intermediate length

$$d^2 = \frac{6}{5}(2\xi)^2.$$

Richter et al [88] and Ewen et al [89, 90] have found the values of the tube diameter for different systems at different temperatures.

The results of the investigations [83 – 90] confirm the existence of the dynamic intermediate length and this is satisfactory from the point of view of all theories. However, it does not mean that investigators confirm the reptation of long macromolecules. The existence of an intermediate length is the effect of first order in respect of co-ordinates in the equation of macromolecule dynamics, while the reptation of a macromolecule is connected to terms of higher order. Other arguments are needed to confirm the reptation mobility.

5.5
The Slowest Relaxation Processes

Conformational relaxation of a macromolecular coil is described by the two relaxation branches (81), while one of them contains large relaxation times, the other small. Meanwhile, there could be a competing mechanism for relaxation of the macromolecular coil to equilibrium due to reptation, so the largest relaxation times, which at $B \gg 1$ can be approximated as

$$\tau_\alpha = \frac{\tau}{2} + \frac{\tau^* B}{\alpha^2}\left(1 + \frac{E}{B}\right),$$

ought to be compared with the reptation relaxation times. The real situations are described by the relation $B \ll E$, so that the relaxation times for very small mode numbers can be approximated by the relation

$$\tau_\alpha = \frac{\tau^* E}{\alpha^2}. \tag{111}$$

According to Doi and Edwards [9, p. 196] the time behaviour of the equilibrium correlation function is described by a formula which is similar to formula for the Rouse chain but the Rouse relaxation times replaced by the reptation relaxation times

$$M_\alpha(t) \sim \exp\left(-\frac{t}{2\tau_\alpha^{\text{rep}}}\right), \quad \tau_\alpha^{\text{rep}} = \frac{\tau_d}{2\alpha^2}, \quad \alpha = 1, 2, \ldots, \ll N. \tag{112}$$

The characteristic reptation relaxation time τ_d in our previous notations can be written as

$$\tau_d = \frac{\zeta b^4 N^3}{\pi^2 \xi^2 T}, \tag{113}$$

In the Doi-Edwards model (Sect. 3.4) we equate the mean square separation between adjacent particles b and the radius of the tube ξ: $b = \xi$. We can use formulae (33) and (90) to express the characteristic time in the form

$$\tau_d = 6\tau^* \frac{\langle R^2 \rangle_0}{\xi^2} = \frac{3}{2} \pi^2 \frac{\tau^*}{\chi}. \tag{114}$$

So, the reptation branch of relaxation times, taking relation (99) into account, can be written as

$$\tau_\alpha^{\text{rep}} = \frac{3}{4} \psi \frac{\tau^*}{\alpha^2}. \tag{115}$$

This branch ought to be compared to the largest times of relaxation of the macromolecule coil which are defined by relation (111). One can see that relaxation times (111) and (115) have a similar dependence on the mode number, but the relaxation times of the reptation branch B times less than relaxation times from the fist conformation branch. The dependence on the length of the macromolecule can be written for two alternative types of motion as

$$\tau_\alpha \sim \begin{cases} M^{3+\delta}, & \textit{first conformation branch} \\ M^3, & \textit{reptation branch} \end{cases}$$

where index δ can be estimated theoretically ($\delta > 2$) and empirically according to the measurements of the characteristics of viscoelasticity ($\delta \approx 2.4$).

So, one ought to conclude that relaxation of the macromolecular coil is realised through reptation instead of the more slow mechanism of rearrangement of all the entangled chains if the parameter $B > 1$ which is always valid for concentrated solutions and melts of polymer. To be consistent, it is necessary to include the anisotropy of motion of the Brownian particles into the dynamic equation (38). One can expect that in properly formulated non-linear dynamics, the branches (111) and (115) will emerge in different limiting cases from one expression for general conformation branch. In any case, one ought to keep into one's mind that both the reptation branch and the branch of largest relaxation times describe large-scale conformational changes. So, we can refer to these branches as to the conformational branches as well.

5.6
Fundamental Dynamical Parameters

The parameters of the theory introduced previously play a fundamental role in the description of the behaviour of a polymer system. So, it is worth-while to discuss them in some detail and to consider their dependencies on the length of the macromolecule and the concentration of polymer in the system.

First of all, we need a scale time of relaxation τ^* which is connected with the coefficient of friction of the Brownian particle in a "monomeric" liquid. According to formula (33)

$$\tau^* \sim \zeta M^2. \tag{116}$$

The scale time of relaxation is a mutual parameter for all existing theories.

To describe the behaviour of a macromolecule in an entangled system, the mesoscopic theory has introduced the correlation time τ and two parameters B and E connected with the external and the internal resistance, respectively.

As we can see in Sect. 5.2, formula (90), the introduced time of correlation appears to be connected with the internal length ξ or (as we shall see in Sect. 6.6, formula (136)) with the length of the macromolecule between adjacent entanglements M_e

$$\chi = \frac{\tau}{2B\tau^*} \approx \frac{(2\xi)^2}{\langle R^2 \rangle} \approx \frac{M_e}{M}.$$

So, one can consider the parameters τ, χ, ξ, and M_e to be equivalent. One of these parameters is introduced in each theory of polymer dynamics. Note that the correlation time is expressed through the two-particle correlation function and the dynamic structure factor of the system of the interacting Brownian particles in many-chain theories [54, 91].

Of course, parameter χ should not be considered as a free parameter. It can be estimated for a given polymer system. We shall consider an entangled polymer system, for which

$$\chi \ll 1. \tag{117}$$

The quantities χ and ψ appear to be connected with each other. Indeed, we have found in Sect. 5.3 that at $B \gg 1$

$$\psi = \pi^2 \frac{1}{\chi}.$$

This formula will be confirmed in Sect. 6.5 in another way. It follows the requirement of the self-consistency of the theory.

Now we shall turn to the discussion of the dependencies of the parameters on the molecular length M and the concentration of polymer c. One can note that the phenomenological parameters B and χ can be written as functions of a single argument. Actually, since the above kinetic restrictions on the motion of a macromolecule are related to the geometry of the system, the only parameters in this case are the number of macromolecules per unit volume n and the mean square end-to-end distance $\langle R^2 \rangle$, while

$$n \sim \frac{c}{M}, \quad \langle R^2 \rangle \sim C_\infty(T)M.$$

The dimensionless quantities B and χ can therefore be regarded as universal and independent of the chemical structure of the polymer, functions of the dimensionless parameter

$$n\langle R^2 \rangle^{3/2} \sim cC_\infty^{3/2}M^{1/2}. \tag{118}$$

To determine the form of the functions, we turn to the known dependencies of the parameters on molecular length, given by formulae (50) and (91)

$$B \sim M^\delta, \ \chi \sim M^{-1}.$$

From the above relations, it follows that

$$\chi \sim c^{-2}C_\infty^{-3}M^{-1}, \ B \sim c^{2\delta}C_\infty^{3\delta}M^\delta. \tag{119}$$

The above-written relations are applicable to all linear polymers, whatever their chemical structure. The relations are valid for molecular length M and concentration c, for which inequality (117) is true. However, the coefficients of proportionality in formulae (119) are the individual characteristics of polymers. They must be estimated empirically. Note that the argument of the functions χ and B can be excluded, so that one has the relation

$$B \sim \chi^{-\delta}.$$

Thus, we can see that the all introduced parameters can be expressed through the only one. The theory contains only one index δ, which has been estimated theoretically in Sect. 3.3 ($\delta = 2$). The experimental estimation gives the value of $\delta = 2.4$ (see Sect. 6.7).

Now the dependencies on the molecular length and concentration for all quantities can be given. For example, formulae (104) for the coefficient of self-diffusion of the macromolecules can be generalised as

$$D \sim \begin{cases} c^{-2\delta}M^{-1-\delta}, & non-reptation \\ c^{-2}M^{-2}, & reptation \end{cases}. \tag{120}$$

One can use formula (90) and (119) to represent the intermediate length as

$$\xi^2 \sim c^{-2}C_\infty^{-2}(T). \tag{121}$$

This relation was confirmed by Kholodenko [92] who started from detailed picture of geometrically confined polymer chain and used more sophisticated methods of calculation.

According to relation (121), the quantity $\xi c C_\infty(T)$ does not depend on the temperature. Nevertheless, Richter et al (1993), at investigating the temperature dependence of the intermediate length in polyethylene-propylene melts, have found that the quantity $\xi c C_\infty(T)$ increases slightly when the temperature increases. One ought to remember that the written relations are only a first approximation.

6
Linear Viscoelasticity

6.1
The Stress Tensor

There are different approaches to the calculation of stresses in the system of entangled macromolecules [9, 34, 93, 94]. We consider the system to be a suspension of Brownian particles and start here to determine a expression for the stress tensor through correlation functions of the normal co-ordinates of the macromolecule, while we shall follow a method developed in the theory of liquids [95, 96] which was used by Pokrovskii and Volkov [55] in this case.

As before, we shall consider each macromolecule either in dilute or in concentrated solution to be schematically represented by a chain of $N + 1$ Brownian particles, so that a set of the equations for motion for the macromolecule can be written as a set of coupled stochastic equations

$$m\frac{d^2 r^\alpha}{dt^2} = F^\alpha + G^\alpha + K^\alpha + \phi^\alpha, \quad \alpha = 0, 1, \ldots, N, \tag{122}$$

where m is the mass of a Brownian particle associated with a piece of the macromolecule of length $M/(N + 1)$, r^α are the co-ordinates of the Brownian particles. The forces acting on the particles was discussing in Sect. 3.1 for dilute solutions and in Sect. 3.2 for entangled systems.

We consider n to be the number density of macromolecular coils in the system, so that the system contains $n(N + 1)$ Brownian particles in unit volume. This number is sufficiently large to introduce macroscopic variables for the suspension of Brownian particles, namely, the mean density

$$\rho(x, t) = \sum_{a,\alpha} m\langle\delta(x - r^{a\alpha})\rangle = m(N + 1)n(x, t), \tag{123}$$

and the mean density of the momentum

$$\rho v_j(x, t) = \sum_{a,\alpha} m\langle u_j^{a\alpha}\delta(x - r^{a\alpha})\rangle. \tag{124}$$

The angle brackets denote averaging over the ensemble of the realisation of random forces in the equations of motion of the particle. The sums in (123) and (124) are evaluated over all the Brownian particles. The double index $a\alpha$ consists of the label of a chain a and the label of a particle α in the chain.

One can start with the definition of the momentum density, given by (124), which is valid for an arbitrary set of Brownian particles. Differentiating (124) with respect to time, one finds

$$\frac{\partial}{\partial t}\rho v_j = -\frac{\partial}{\partial x_i}\sum_{a,\alpha} m\langle u_i^{a\alpha} u_j^{a\alpha}\delta(x - r^{a\alpha})\rangle + \sum_{a,\alpha}\langle m\frac{du_j^{a\alpha}}{dt}\delta(x - r^{a\alpha})\rangle. \qquad (125)$$

The right-hand side of equation (125) has to be reduced to a divergent form. To transform the second term into the required form, we use the dynamic equation (122) which, in this general form, is valid for a macromolecule both in a viscous liquid and in an entangled system. After summing over all the particles of the macromolecule and averaging, one can write for each macromolecular coil

$$m\sum_{\alpha=0}^{N}\langle\frac{du^{a\alpha}}{dt}\delta(x - r^{a\alpha})\rangle = \sum_{\alpha=0}^{N}\langle(K^{a\alpha} + G^{a\alpha})\delta(x - r^{a\alpha})\rangle, \quad a = 1, 2, \ldots, n.$$

whereas the requirement that there is no mean volume force is used

$$\sum_{\alpha=0}^{N}\langle(F^{a\alpha} + \phi^{a\alpha})\delta(x - r^{a\alpha})\rangle = 0, \quad a = 1, 2, \ldots, n.$$

Next, the formal expansion of the δ-function into a Taylor's series about the centre of mass q^a of the ath macromolecule can be used, retaining only the first two terms of the expansion

$$\delta(x - r^{a\alpha}) = \delta(x - q^a) - (r_k^{a\alpha} - q_k^a)\frac{\partial}{\partial x_k}\delta(x - q^a).$$

So, the above formula is transformed into

$$-\frac{\partial}{\partial x_k}\sum_{\alpha=0}^{N}\langle(K_j^{a\alpha}r_k^{a\alpha} + G_j^{a\alpha}r_k^{a\alpha})\delta(x - q^a)\rangle, \quad a = 1, 2, \ldots, n.$$

Here, the sum is conducted over all the particles in a chosen macromolecule. Assuming that all the macromolecules are identical, and neglecting the statistical dependence of the position of the centres of mass of the macromolecules on the other co-ordinates, one obtains an expression for the second term on the right-hand side of Eq. (125) in the divergent form

$$\sum_{a,\alpha}\langle m\frac{du_j^{a\alpha}}{dt}\delta(x - r^{a\alpha})\rangle = -\frac{\partial}{\partial x_k}n\sum_{\alpha=0}^{N}\langle K_j^\alpha r_k^\alpha + G_j^\alpha r_k^\alpha\rangle.$$

The first term on the right-hand side of (125) can also be rewritten in a more convenient form. One uses the definition of the mean velocity v_i and, taking only the first term of the expansion of the δ-function into account, one finds that

$$m \sum_{a,\alpha} \langle u_j^{a\alpha} u_i^{a\alpha} \delta(x - q^a) \rangle = nm \sum_{\alpha=0}^{N} \langle (u_j^\alpha - v_j)(u_i^\alpha - v_i) \rangle + \rho v_i v_j.$$

Thus, an equation which has the sense of a law of conservation of momentum has been obtained. There is an expression for the momentum flux $\rho v_i v_j - \sigma_{ij}$ under the derivation symbol, which allows one to write down the expression for the stress tensor

$$\sigma_{kj} = -n \sum_{\alpha=0}^{N} \left[m \langle (u_j^\alpha - v_j)(u_k^\alpha - v_k) \rangle + \langle K_k^\alpha r_j^\alpha + G_k^\alpha r_j^\alpha \rangle \right].$$

The assumption that the particle velocities are described by the local-equilibrium distribution yields

$$\sigma_{ik} = -n(N + 1)T\delta_{ik} - n \sum_{\alpha=0}^{N} \langle K_i^\alpha r_k^\alpha + G_i^\alpha r_k^\alpha \rangle. \tag{126}$$

In this form the expression for the stresses is valid for any dynamics of the chain. For the macromolecules in an entangled system, the elastic and internal viscosity forces according to equations (26) and (40) – (42) have the form

$$K_i^\alpha = -2T\mu A_{a\gamma} r_i^\alpha, \quad G_i^\alpha = - \int_0^\infty G_{a\gamma}\, \varphi(s)(u_i^\gamma - \omega_{ij} r_j^\gamma)_{t-s}\, ds.$$

It is possible now to write the stress tensor in the following form in terms of normal co-ordinates

$$\sigma_{ik}(t) = -n(N + 1)T\delta_{ik} + nT \sum_{\alpha=1}^{N} \Big\{ 2\mu\lambda_\alpha \langle \rho_i^\alpha \rho_k^\alpha \rangle - \delta_{ik} \tag{127}$$

$$+ \frac{1}{T} \int_0^\infty \varphi(s) \Big(\langle \dot\rho_i^\alpha(t-s)\rho_k^\alpha(t) \rangle - \omega_{il}(t-s)\langle \rho_i^\alpha(t-s)\rho_k^\alpha(t) \rangle \Big) ds \Big\}.$$

The stress tensor for the "monomer" liquid can be added to expression (127). However, the contribution of the viscoelastic carrier in the case of a concentrated solution is slight, and we shall ignore it.

6.2
Dynamic Modulus and Relaxation Branches

To calculate stresses for a system of weakly coupled macromolecules, non-equilibrium correlation functions (71), (74) and (75), specified for the memory function (45), can be used to write down the stresses in linear approximation with respect to the velocity gradients. In this way, the stresses are determined by the velocity gradients in all the previous moments of time. Further on, we will consider the viscoelastic behaviour of the polymer system when a shear oscillatory motion is imposed

$$\gamma_{ik} \sim e^{-i\omega t}.$$

The expression for stresses (127) then assumes the form

$$\sigma_{ik}(t) = -p\delta_{ik} + 2\eta(\omega)\gamma_{ik}(t) \tag{128}$$

where pressure p includes the partial pressures of the monomer liquid and the liquid of Brownian particles. Expression (128) defines the complex viscosity coefficient

$$\eta(\omega) = \eta'(\omega) + i\eta''(\omega).$$

It is also convenient to describe the reaction of the system in terms of the dynamic modulus which is related to the dynamic viscosity by

$$G(\omega) = G'(\omega) - iG''(\omega) = -i\omega\eta(\omega).$$

To begin with, let us consider the case of very low frequencies, so that ζ can be neglected in comparison to ζB and ζB in the correlation functions. In this case, the stress tensor (127) gives us an expression for the dynamic modulus which after some rearrangement can be written in the standard form

$$G(\omega) = nT \sum_{\alpha=1}^{N} \left(\frac{\tau_\alpha^R}{\tau_\alpha}\right)^2 B\left(B\frac{-i\omega\tau_\alpha}{1 - i\omega\tau_\alpha} + E\frac{\tau_\alpha}{\tau}\frac{-i\omega\tau_\alpha^*}{1 - i\omega\tau_\alpha^*}\right). \tag{129}$$

where

$$\tau_\alpha = \frac{\tau}{2} + \tau_\alpha^R(B + E), \quad \tau_\alpha^* = \frac{2\tau\tau_\alpha}{2\tau_\alpha + \tau}, \quad \tau_\alpha^R = \tau^*/\alpha^2 \tag{130}$$

We can, thus, see that, at low frequencies, the viscoelastic behaviour of the system is determined by two sets of relaxation times in which, for small α and large B, we have

$$\tau_\alpha \gg \tau_\alpha^*, \quad \tau_\alpha^* \approx \tau.$$

We can say also that the viscoelastic behaviour is determined by two relaxation branches. Note that the first and the second terms in (129) at $\omega \to \infty$ have the orders of magnitudes nT and $nT\chi^{-1}$ respectively. So, the contribution of the first, conformation branch to the linear viscoelasticity is small.

To extend the theory to higher frequencies, we have to use the entire forms of correlation functions and, after some rearrangement, we find the dynamic modulus

$$G(\omega) = nT \sum_{a=1}^{5} \sum_{\alpha=1}^{N} (-i\omega) \frac{p_\alpha^{(a)}\tau_\alpha^{(a)}}{1 - i\omega\tau_\alpha^{(a)}} \tag{131}$$

where the times of relaxation and the corresponding weights are given by the following expressions

$$2\tau_\alpha^{(1)} = 2\tau_\alpha^+ = \tau_\alpha + \left(\tau_\alpha^2 - 2\tau\tau_\alpha^R\right)^{1/2}, \quad \tau_\alpha^{(2)} = \frac{2\tau\tau_\alpha^+}{\tau + 2\tau_\alpha^+},$$

$$\tau_\alpha^{(3)} = \frac{2\tau_\alpha^+\tau_\alpha^-}{\tau_\alpha^+ + \tau_\alpha^-}, \quad \tau_\alpha^{(4)} = \frac{2\tau\tau_\alpha^-}{\tau + 2\tau_\alpha^-}, \quad 2\tau_\alpha^{(5)} = 2\tau_\alpha^- = \tau_\alpha - \left(\tau_\alpha^2 - 2\tau\tau_\alpha^R\right)^{1/2}$$

$$p_\alpha^{(1)} = T_\alpha^+ S_\alpha^+ \left(1 - \frac{2E\tau_\alpha^R}{2\tau_\alpha^+ - \tau}\right)$$

$$p_\alpha^{(2)} = S_\alpha^+ \frac{E\tau_\alpha^R}{\tau(\tau_\alpha^+ - \tau_\alpha^-)} + T_\alpha^+ S_\alpha^+ \frac{2E\tau_\alpha^R}{2\tau_\alpha^+ - \tau} - (T_\alpha^+ S_\alpha^- + T_\alpha^- S_\alpha^+) \frac{2E\tau_\alpha^R}{2\tau_\alpha^+ - \tau}$$

$$p_\alpha^{(3)} = (T_\alpha^+ S_\alpha^- + T_\alpha^- S_\alpha^+) \left(\frac{E\tau_\alpha^R}{2\tau_\alpha^+ - \tau} + \frac{E\tau_\alpha^R}{2\tau_\alpha^- - \tau} - 1\right)$$

$$p_\alpha^{(4)} = -S_\alpha^- \frac{E\tau_\alpha^R}{\tau(\tau_\alpha^+ - \tau_\alpha^-)} + T_\alpha^- S_\alpha^- \frac{2E\tau_\alpha^R}{2\tau_\alpha^- \tau} - (T_\alpha^+ S_\alpha^- + T_\alpha^- S_\alpha^+) \frac{2E\tau_\alpha^R}{2\tau_\alpha^- - \tau}$$

$$p_\alpha^{(5)} = T_\alpha^- S_\alpha^- \left(1 - \frac{2E\tau_\alpha^R}{2\tau_\alpha^- - \tau}\right).$$

Expression (131) generalises formula (129) and describes the frequency dependence of the dynamic modulus at higher frequencies. The dynamic modulus is determined by five relaxation branches in this case.

Figure 3 shows the calculated and measured dynamic shear modulus of polystyrene, for values of χ and B, chosen so as to ensure agreement with the modulus on the plateau and the length of the plateau. Good agreement is achieved at high and low frequencies. The form of the real part of the modulus confirms the assumption about the finite correlation time τ in the memory functions (45), but discrepancies in the region of the plateau (especially seen on the $G''(\omega)$ plot) witness that other correlation times in the memory function less than τ can exist. In fact, to describe the dependencies for monodisperse polybutadiene and polystyrene empirically, a few relaxation times were introduced [98, 99]. However, the differences in the region of the plateau can also be attributed to the inevitable polydispersity of the samples.

The contributions of the relaxation branches are also depicted in Fig. 3. The slowest of the relaxation branches (conformation relaxation) is practically absent from these plots. The first branch changes the values of the imaginary part of the modulus insufficiently, and a step appears in the frequency dependence of the real part of the dynamic modulus. The most essential contribution is given by the second relaxation branch with relaxation times close to τ. The remaining branches merge together and form a group of slow relaxation times, so that two groups are usually noted [100], namely, slow and fast. This picture is typical for concentrated polymer solutions and melts. The recent examples of the dependencies for polybutadiene and for polystyrene can be found in papers by Baumgaertel et al [98, 99].

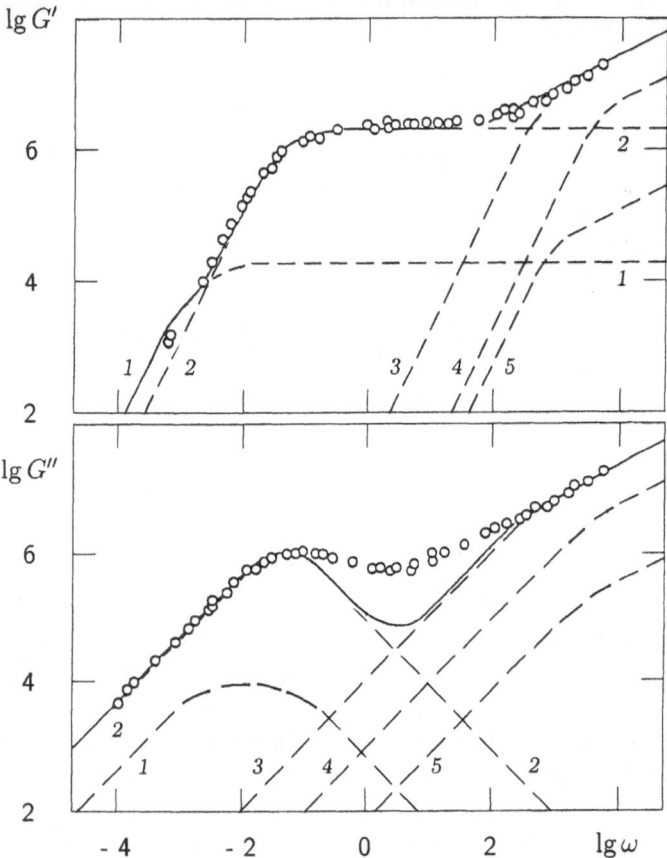

Fig. 3. Dynamic shear modulus of polystyrene: experimental points and theoretical de-pendence. The separate contributions from relaxation branches are shown by *dashed curves*: 1 - conformational branch; 2 - orientation or viscoelastic branch; 3, 4, 5 - glassy branches. The points are taken from work by Onogi et al [97] for polystyrene with mole-cular weight 215 000 at $T = 160\,°C$. The theoretical dependencies were calculated for the values: $B = 3000$, $E = 20, 000$, $\chi = 0.08$, $\tau^* = 5 \times 10^{-5}$ s, $nT = 1.7 \times 10^4$ Pa. Adapted from the paper of Pokrovskii [11]

6.3
Self-Consistency of the Theory

One can notice that the dissipative terms in the dynamic equation (41) (taken for the case of zero velocity gradients, $\nu_{ij} = 0$) have the form of the resistance force (A.3) for a particle moving in a viscoelastic liquid, while the memory functions are (with approximation to the numerical factor) fading memory functions of the viscoelastic liquid. The macromolecule can be considered as moving in a viscoelastic continuum. In the case of choice of memory functions (45), the medium has a single relaxation time and is characterised by the dynamic modulus

$$G(\omega) = \frac{-i\omega\eta_m}{1 - i\omega\tau}, \qquad G_e = \lim_{\omega\to\infty} G(\omega) = \frac{\eta_m}{\tau},$$

where τ is the correlation time, and η_m is a constant coefficient of microviscosity. One can say that the written dynamic modulus characterises the micro-viscoelasticity.

On the other hand, the properties of the system as a whole can be calculated and the macroscopic dynamic modulus can be determined. Here the question of the relation between the postulated micro-viscoelasticity and the resulting macro-viscoelasticity appears. The answer requires a properly formulated self-consistency condition. Simple speculations show that equality of the micro- and macro-viscoelasticity cannot be obtained. Nevertheless, it is natural to require the equality of relaxation times of micro- and macro-viscoelasticities. It will be shown in this Subsection that this condition can be satisfied.

First of all, we shall consider in detail the characteristic quantities. Expansion $G(\omega) = -i\omega\eta + \omega^2\nu$ determines the viscosity coefficient η and the elasticity coefficient ν. In addition, the value of the dynamic modulus on the plateau G_e has to be considered as an essential characteristic of viscoelasticity. It can be calculated as the limiting value of the modulus at the intermediate frequencies. One can use expression (129), to evaluate the described quantities

$$G_e = nT \sum_{\alpha=1}^{N} \frac{2\chi\alpha^2 + \psi(\chi\alpha^2 + 1 + \psi)}{2\chi\alpha^2(\chi\alpha^2 + 1 + \psi)^2}$$

$$\eta = nT\tau^*B \sum_{\alpha=1}^{N} \left[\frac{1}{\alpha^2(\chi\alpha^2 + 1 + \psi)} + \frac{\psi}{\alpha^2(2\chi\alpha^2 + 1 + \psi)} \right]$$

$$\nu = nT(\tau^*B)^2 \sum_{\alpha=1}^{N} \frac{2\chi\psi(\chi\alpha^2 + 1 + \psi)}{\alpha^2(2\chi\alpha^2 + 1 + \psi)}.$$

A preliminary estimate of χ which, according to (91), can be interpreted as the ratio of the square of the tube diameter $(2\xi)^2$ to the mean square end-to-end distance $\langle R^2 \rangle_0$, shows that $\chi \ll 1$ for sufficiently long macromolecules. For large N, this enables us to replace summation by integration and to obtain expressions for the characteristic quantities

$$G_e = nT \left[\frac{\pi^2}{12} \frac{\psi}{\chi(1+\psi)} + \frac{\pi}{8} \frac{2-\psi}{(1+\psi)^{3/2}} \chi^{-1/2} \right]$$

$$\eta = nT\tau^*B \left[\frac{\pi^2}{6} - \frac{\pi}{2} \left(\frac{\chi}{1+\psi} \right)^{1/2} \right]$$

$$\nu = nT(\tau^*B)^2 \left[\frac{\pi^2}{3} \frac{\chi\psi}{1+\psi} - \frac{\pi}{2} \left(\frac{2\chi}{1+\psi} \right)^{3/2} 2\psi \right]. \tag{132}$$

These expressions are valid for arbitrary ψ and small χ. However, according to relation (99), for systems consisting of very long molecules in the almost complete absence of the solvent we have $\psi \gg 1$. So, one finds the zeroth-order terms in power of ψ^{-1}

$$G_e = \frac{\pi^2}{12} nT\chi^{-1}, \quad \eta = \frac{\pi^2}{6} nT\tau^* B, \quad \nu = \frac{\pi^2}{3} nT(\tau^* B)^2 \chi. \tag{133}$$

We note that the relaxation times of the second branch are very close to each other, so that the frequency dependence of the modulus could be approximated by a expression with the single relaxation time determined by the relation

$$\bar{\tau} = \frac{\eta}{G_e} = 2\tau^* B\chi = \tau.$$

The relaxation time that we have determined may be referred to as the main viscoelastic relaxation time; it is equal to the correlation time which was introduced to characterise the medium surrounding the chosen macromolecule. Thus, for $\psi \to \infty$, the theory is self-consistent. For the first-order terms in power of ψ^{-1} of expansion of (132), the condition of self-consistency, as a requirement of the identity of the times of relaxation of macro- and micro-viscoelasticity, gives the following relation

$$\psi = \frac{4\pi^2}{9} \frac{1}{\chi}. \tag{134}$$

This relation is practically identical to relation (99).

6.4
Modulus of Elasticity and the Intermediate Length

Initially, the elasticity of concentrated polymer systems was ascribed to the existence of a network in the system formed by long macromolecules with junction sites [100]. The sites were assumed to exist for an appreciable time, so that, for observable times which are less than the lifetime of the sites, the system appears to be elastic. The number of entanglements for a single macromolecule M/M_e can be calculated according to the formula

$$G_e = nT\frac{M}{M_e}. \tag{135}$$

The length of a macromolecule between neighbouring entanglements M_e was used as a characteristic of the polymer system. A list of estimates of M_e can be found in work by Aharoni [101, 102].

Now we can return to expression (133) for the value of the modulus on the plateau and compare it to equation (135). In this way, we can find an expression for the parameter of the theory

$$\chi = \frac{\pi^2}{12} \frac{M_e}{M}. \tag{136}$$

We should note, recalling the interpretation of χ as the ratio of the intermediate length to the size of the coil (formula (91)) discussed previously, that the length M_e, determined by formula (135), is actually related to the intermediate length ξ. Expression (135) can be rewritten in a form which is identical to the relation by Doi and Edwards [9]

$$G_e = \frac{2}{3} nT \frac{\langle R^2 \rangle}{(2\xi)^2}.$$

Note that the squared diameter of the Doi-Edwards tube relates to our intermediate length as follows

$$d^2 = \frac{6}{5} (2\xi)^2$$

The intermediate length (tube diameter) 2ξ can be estimated from the modulus with the aid of the above equations. Comparison of values of the intermediate length found from dynamic modulus and from neutron-scattering experiments was presented by Ewen and Richter [89]. They found the values to be close to each other, though there is a difference in the temperature dependence of the values of intermediate length found by different methods.

Although a network is not present in a concentrated solution, there exists a characteristic length, which had earlier been assumed to be the distance between neighbouring network sites. The characteristic length is a dynamic one. There are no temporary knots in a polymer system, though there is a characteristic time, which is the lifetime of the frozen large-scale conformation of a macromolecule in the system. So, the conceptions of intermediate length and characteristic time are based on deeper ideas and are reflected in the theory.

6.5
Dependence on Concentration and Macromolecular Length

Thus in the mesoscopic approximation or, in other words, in the mean-field approximation, the dynamic shear modulus of the melt or the concentrated solution of the polymer is represented by a function of a small number of parameters

$$G(\omega) = nTf(\tau^*\omega, B, \chi). \tag{137}$$

In the limiting case of very concentrated solutions and melts of high-molecular polymers, we assume that $B \gg 1$. Hence it follows that $\tau > \tau^*$, which fact imposes certain restrictions on χ, so that $1/B < \chi \ll 1$. For these values of B and χ, the theory is found to be self-consistent for $\psi \gg 1$, so that once again, as was shown in Sect. 6.3, the formulae for the dynamic modulus lead to expressions for the characteristic quantities:

$$\eta = \frac{\pi^2}{6} nT\tau^* B, \quad \tau = 2\tau^* B\chi,$$

$$\nu = \frac{\pi^2}{3} nT(\tau^* B)^2 \chi, \quad G_e = \frac{\pi^2}{12} nT\chi^{-1}. \tag{138}$$

Experiments reveal that the dynamic modulus and the characteristic quantities (138) depend on the polymer concentration c and length M of the macromolecule [100].

According to the results of Sect. 5.6, the phenomenological parameters χ and B can be written as functions of a single argument $C_\infty^{3/2} cM^{1/2}$, where C_∞ is the thermodynamic rigidity of the macromolecule. Using these results, one can write the dependencies of the characteristic quantities on the concentration of polymer, and on the thermodynamic rigidity and molecular weight of the macromolecule as

$$\eta \sim \zeta_0 C_\infty^{3\delta+1} c^{2\delta+1} M^{\delta+1}, \qquad \tau \sim \frac{1}{T}\zeta_0 C_\infty^{3\delta-2} c^{2\delta-2} M^{\delta+1},$$

$$\nu \sim \zeta_0^2 C_\infty^{3\delta-1} c^{4\delta-1} M^{2\delta+2}, \qquad G_e \sim T C_\infty^3 c^3 M^0. \tag{139}$$

These dependencies can be compared with experimental determinations. Many data obtained for almost monodisperse samples of polymer melts of different molecular weight show that for high molecular weights, $\eta \sim M^{\delta+1}$, $G_e \sim M^0$, where, as a rule, $\delta = 2.4$ [100, 103]. Empirical estimation differs from the coarse theoretical estimation in Section 3.2, according to which $\delta = 2$. The behaviour of the initial viscosity is the most widely studied [104, 105]. The concentration dependence of the viscosity coefficient in the "melt-like" region can be represented by a power law [106]. The index can be found to be approximately $2\delta + 1$, in accordance with (139). However, there are some differences in the behaviour of polymer solutions, which are connected with different behaviour of macromolecular coils at dilution.

Equation (138) allow one to establish relationships between the functions of the characteristic quantities. For example, according to the experimental data [100], the dependence of η and of τ on the length of a macromolecule is the same.

We should note once again that the above discussion and expressions are valid only for very long macromolecules and in the limit of very high concentrations. For semi-dilute solutions, the analysis should also include another dimensionless parameter (see Sects. 2.4 and 2.5), but then the results would become more complicated.

6.6
Frequency-Temperature Superposition

The dependence of the characteristic quantities (138) on temperature is mainly determined by the monomer friction coefficient ζ_0 which is a function of temperature, concentration, and (for small M) of molecule length [103].

The coefficient of friction is caused by the motion of small portions of the macromolecule, so that its temperature dependence is similar to that found for low-molecular-weight liquids, and can be written in the following form at temperatures much higher than the glass transition point

$$\zeta_0 \sim \exp \frac{U}{T} \tag{140}$$

where U is the activation energy that depends on the molecular weight (for small M), on the concentration, and also on the temperature if the temperature range in which the viscosity is considered is large. Near the glass transition point T_g, we have

$$\zeta_0 \sim \exp \left[\frac{A}{f_g - \alpha(T - T_g)} \right] \tag{141}$$

where A is an individual parameter, f_g is the volume fraction of free volume, and α is the expansion coefficient of the liquid. Quantities A and f_g are practically independent of the concentration and molecular weight, so that the dependence of ζ_0 on c and M is determined by the dependence of T_g on these quantities.

We note that, since the parameters B and χ are practically independent of temperature, the shape of the curves showing G/nT as a function of the dimensionless frequency $\tau^* \omega$ does not change as the temperature increases, so that we can make a superposition using a reduction coefficient obtained from the temperature dependence of the viscosity.

To determine the procedure for the reduction, we shall write down the dynamic modulus at two different temperatures, one of which is a reference temperature T_{ref}

$$G(\omega, T_{\text{ref}}) = nT_0 f(\tau^*_{T_{\text{ref}}} \omega, B, \chi),$$
$$G(\omega, T) = nT f(\tau^*_T \omega, B, \chi).$$

One can consider the parameters B and χ to be independent of the temperature and replace the argument in the first line in such a way as to exclude the dimensionless function. Then we write down the rule for reduction as

$$G(a_T \omega, T_{\text{ref}}) = \frac{\rho_{T_{\text{ref}}} T_{\text{ref}}}{\rho_T T} G(\omega, T), \tag{142}$$

where the shift coefficient is given by

$$a_T = \frac{\tau_T^*}{\tau_{T_{\text{ref}}}^*} = \frac{T_{\text{ref}}\,(C_\infty^{3\delta}\rho^{2\delta+1})_{T_{\text{ref}}}}{T\,(C_\infty^{3\delta}\rho^{2\delta+1})_T}\,\frac{\eta_T}{\eta_{T_{\text{ref}}}}.$$ (143)

The above expressions confirm the known [100] method of reducing the dynamic modulus measured at different temperatures to an arbitrarily chosen standard temperature T_{ref}, while offering a relatively insignificant improvement on the usual shift coefficient

$$a_T = \frac{T_{\text{ref}}\,\rho_{T_{\text{ref}}}\,\eta_T}{T\,\rho_T\,\eta_{T_{\text{ref}}}}.$$

6.7
Viscoelasticity of Dilute Blends of Polymers

By studying a mixture of two polymers, one of which is present in much smaller amounts, one has a unique opportunity to obtain direct information about the dynamics of a chosen single macromolecule in a viscoelastic liquid consisting of the matrix macromolecules [59].

Consider a linear polymer with molecular weight M_0 and a small impurity of a similar polymer with a higher molecular weight M. We shall assume that the amount of the high-molecular-weight impurity is so small that its molecules do not interact with each other, so that the medium in which the molecules move is a system consisting of a linear polymer of molecular weight M_0, which is characterised by the modulus

$$G_0(\omega) = -i\omega\eta_0(\omega).$$

The change in the stress produced by the small amount of macromolecules of another kind is, clearly, determined by the dynamics of the non-interacting impurity macromolecules among the macromolecules of another length, so that this case is of particular interest from the standpoint of the theory of the viscoelasticity of linear polymers.

We shall now turn to the case of low frequencies for which the dynamic modulus can be written in the form of the expansion $G(\omega) = -i\omega\eta + \omega^2\nu$ which determines the viscosity coefficient η and the elasticity coefficient ν.

We shall begin by considering the characteristic quantities

$$[\eta] = \lim_{c\to 0}\frac{\eta - \eta_0}{c\eta_0}, \quad [\nu] = \lim_{c\to 0}\frac{\nu - \nu_0}{c\nu_0},$$ (144)

as functions of the length (or molecular weight) of the macromolecules of the matrix and the impurity. The index 0 refers to the matrix and c is the impurity concentration.

To calculate the quantities $\eta - \eta_0$ and $\nu - \nu_0$, we use Eq. (133) given in Sect. 6.3. We should note that for the considered blends of polymers, we must consider B, E and τ as functions of M_0, and τ^* as a function of M, so that, for the macromolecules of the matrix and the additive, we now have, respectively,

$$\chi_0 \sim M_0^{-1}, \quad \chi \sim M_0 M^{-2}. \tag{145}$$

We shall discuss the case when these two quantities are small, which ensures that the formulae given by (133) are valid both for the matrix and for the additive. Now, we ought to take into account the fact that some of the macromolecules of the matrix have been replaced by impurity macromolecules.

Actually, a system that contains n_0 matrix macromolecules and n impurity macromolecules per unit volume can be characterised by assuming the additivity by dynamic modulus

$$G - G_0 = n\left(g - \frac{M}{M_0} g_0\right)$$

where g and g_0 are the contributions to the dynamic modulus from a single macromolecule of the impurity and the matrix, respectively. Taking this into account, we can rewrite the expressions for the characteristic quantities in the form

$$\eta - \eta_0 = \frac{\pi^2}{6} n T \tau^* B\left(1 - \frac{M_0}{M}\right), \quad \nu - \nu_0 = \frac{\pi^2}{3} n T (\tau^* B)^2 \chi\left(1 - \frac{M_0}{M}\right). \tag{146}$$

When $M \gg M_0$, the above expressions reduce to (133). From formula (146) we find expressions for the characteristic quantities

$$[\eta] = \frac{\pi^2}{6} \frac{n T \tau^* B}{c \eta_0}\left(1 - \frac{M_0}{M}\right), \quad [\nu] = \frac{\pi^2}{3} \frac{n T (\tau^* B)^2 \chi}{c \nu_0}\left(1 - \frac{M_0}{M}\right). \tag{147}$$

We shall assume that the length of the macromolecules in the matrix is such that the discussion and results of Sect. 6.5 stand, so that

$$\eta_0 = \frac{\pi^2}{6} n_0 T \tau_0^* B, \quad \nu_0 = \frac{\pi^2}{3} n_0 T (\tau_0^* B)^2 \chi_0$$

where n_0 is the number of the matrix macromolecules per unit volume and τ_0^* is the characteristic relaxation time of the macromolecules of the matrix.

Using the above relations and equations (147), we find that for $M \gg M_0$

$$[\eta] \sim M_0^{-1} M, \quad [\nu] \sim M_0^{-1} M. \tag{148}$$

On the other hand, when $M \ll M_0$, the characteristic quantities are negative and are independent of the length of the matrix and of the impurity macromolecules

$$[\eta] \sim M_0^0 M^0, \quad [\nu] \sim M_0^0 M^0. \tag{149}$$

Results (148) and (149) do not depend upon any dependence of B on the length (molecular weight) of the macromolecule.

The viscoelastic behaviour of dilute blends of polymers of different length and narrow molecular weight distributions was investigated experimentally for polybutadiene by Yanovski et al. [107] and by Jackson and Winter [108] and for polystyrene by Watanabe et al [109, 110] (the results can be found in the work by Jackson and Winter [108]). The results for polybutadiene was approximated by Pokrovskii and Kokorin [59] by the dependencies

$$[\eta] \sim M_0^{-0.8} M^{0.5}, \quad [\nu] \sim M_0^{-(1.8 \to 2.2)} M^{1.3 \to 3.0}. \tag{150}$$

Though it is desirable to confirm the experimental results, the comparison of the theoretical formulas (148) with the experimental results (150) indicates that, insofar as the influence of the ambient macromolecules on the dynamics of a chosen macromolecule is concerned, the ambient macromolecules are equivalent to a certain relaxing medium. The reptation effect is due to terms of order higher than the first in the equation of motion of the macromolecule, and it is actually the first-order terms that dominate the linear viscoelastic phenomena. Attempts to describe viscoelasticity without the leading linear terms lead to a distorted picture, so that one begins to understand the lack of success of the reptation model in the description of the viscoelasticity of polymers. Reptation has to be included when one considers the non-linear effects in viscoelasticity.

7
Optical Anisotropy

7.1
The Relative Permittivity Tensor

The macromolecular coils are deformed in flow, while optically anisotropic parts of the macromolecules are oriented by flow, so that polymers and their solutions become optically anisotropic.

To consider the relative permittivity tensor of polymeric system, one makes use of the heuristic model of a macromolecule as freely-jointed segments: each macromolecule consists of z segments and is surrounded by solvent molecules (Sect. 2.1). Then, the simple old-fashion [111, 112] speculations allow us to determine the relative permittivity tensor of polymeric system in terms of the mean orientation of anisotropic segments of the macromolecules $\langle e_i e_k \rangle$. The relative permittivity tensor is formulated below to within first-order terms in the orientation tensor

$$\varepsilon_{ik} = \varepsilon_0 \delta_{ik} + 4\pi n \, z \Delta\alpha \left(\frac{\varepsilon_0 + 2}{3}\right)^2 \left(\langle e_i e_k \rangle - \frac{1}{3}\delta_{ik}\right) \tag{151}$$

where n is the density of the number of macromolecules in the system, z is the number of Kuhn segments in the macromolecule and $\Delta\alpha$ is the anisotropy of the polarisability of a Kunn segment.

The relative permittivity tensor can be also written with greater accuracy [45, 129]. In the second approximation, the principal axes of the relative permittivity tensor do not coincide, generally speaking, with the principal axes of the orientation tensor. It is readily seen that interesting situations may arise when $\Delta\alpha < 0$; in this case, the coefficients of the first- and second-order terms have different signs. The experimental data were discussed by Gotlib and Svetlov [113, 114].

Now, we have to return to the subchain model of macromolecule, which was used to calculate the stresses in the polymeric system, and express the tensor of the mean orientation of the segments of the macromolecule in terms of the subchain model.

7.1.1
Dilute Solution

In the equilibrium situation, at a given end-to-end distance R of a macromolecule, the tensor of mean orientation of segments of a chain is determined [13] as

$$\langle e_i e_k \rangle - \frac{1}{3}\delta_{ik} = \frac{3}{5(zl)^2}\left(R_i R_k - \frac{1}{3}R^2 \delta_{ik} \right). \tag{152}$$

As before, we shall consider each macromolecule to be divided into N subchains and assume that every subchain of the macromolecule is in the equilibrium. So, using the above formula relating the tensor of the mean orientation of the segments of the macromolecules to the distance between the ends of the subchains, we arrive from relation (151) at Zimm's [5] expression for the relative permittivity tensor

$$\varepsilon_{ik} = \varepsilon_0 \delta_{ik} + n\Gamma\left(\langle r_i^\alpha A_{\alpha\gamma} r_k^\gamma \rangle - \frac{1}{3}\langle r_j^\alpha A_{\alpha\gamma} r_j^\gamma \rangle \delta_{ik} \right) \tag{153}$$

which in terms of the normal co-ordinates introduced by means of Eqs. (12), assumes the form

$$\varepsilon_{ik} = \varepsilon_0 \delta_{ik} + n\Gamma \sum_{\alpha=1}^{N} \lambda_\alpha\left(\langle \rho_i^\alpha \rho_k^\alpha \rangle - \frac{1}{3}\langle \rho_j^\alpha \rho_j^\alpha \rangle \delta_{ik} \right). \tag{154}$$

In these formulae, n is the density of the number of macromolecules in the solution, while the coefficient of the anisotropy of the macromolecular coil Γ is given by the following expression in the case where the macromolecule is modelled by a freely-jointed chain of Kuhn segments

$$\Gamma = 4\pi\Delta\alpha\left(\frac{\varepsilon_0 + 2}{3} \right)^2 \frac{3N}{5zl^2},$$

where z is the number of Kuhn segments in the macromolecule, and $\Delta\alpha$ is the anisotropy of the polarisability of a Kunn segment.

The anisotropy of the macromolecular coil Γ is characteristic of a macromolecule independent of any model used for calculation. One must also take into account the possible effect of the shielding of the inner segments of the macromolecular coil, the latter effect being the greater the longer the macromolecule, so that the expression for the anisotropy coefficient, which must be covariant in relation to subdivisions into subchain, assumes the form

$$\Gamma = 4\pi\Delta\alpha\left(\frac{\varepsilon_0 + 2}{3}\right)^2 \frac{3N^{2\nu}}{5\langle R^2\rangle}. \tag{155}$$

The anisotropy of the macromolecule has been calculated for other chain models. Expressions for the anisotropy coefficient are known in the case where the macromolecule has been represented schematically by a continuous thread (the persistence length model) [115, 116] and also in the case where the microstructure of the macromolecules has been specified. In the latter case, the anisotropy coefficient of the macromolecule is expressed in terms of the bond polarisabilities and other microcharacteristics of the macromolecule [13].

Expression (153) for the relative permittivity tensor is valid only to a first approximation as regards the orientation of the segments and describes the anisotropy of the system associated with the intrinsic anisotropy of the segments. It was assumed that distribution of orientation of the segments inside every subchain are considered to be in equilibrium though under deformation which is valid for low frequencies. Nevertheless, expression (153) for the relative permittivity tensor has appeared to be very well applicable to dilute polymer solutions [117, 118]. To discuss the problem for the higher frequencies, one has to introduce in equation (152) terms describing deviation of the mean orientation of segments from equilibrium under deformation.

7.1.2
Entangled System

The situation is different for very concentrated polymer solutions. Though equation (151) is applicable for this case, formula (152) is not valid neither for the entire macromolecule nor for a separate subchain. The subchain of a macromolecule in the deformed entangled system is not in equilibrium even in the fist approximation, and the problem about distribution of orientation of the interacting, connected in chains, segments apparently is not solved yet.

In this situation we refer to experimental evidence according to which components of the relative permittivity tensor are strongly related to components of the stress tensor. It is usually stated [9] that the stress-optical law, that is proportionality between the tensor of relative permittibility and the stress tensor, is valid for an entangled polymer system, though one can see (for example, in some plots of the paper by Kannon and Kornfield [119]) deviations from the stress-optical law in the region of very low frequencies for some samples. In linear approximation for the region of low frequencies, one can write the following relation

$$\varepsilon_{ij} - \varepsilon_0 \delta_{ij} = 2\bar{n} C \left(\sigma_{ij} + p\delta_{ij} \right), \tag{156}$$

where $\bar{n}$ is an isotropic value of the refractive index ($\bar{n}^2 = \varepsilon_0$) and C is the stress-optical coefficient which in molecular terms for low frequencies can be written as

$$C = \frac{\Gamma}{4\pi\mu T}$$

Relation (156) reflects the fact that both the stresses and the optical anisotropy of a polymeric liquid under motion are determined by the mean orientation of the interacting segments. One can use expression (127) for the stress tensor to write

$$\varepsilon_{ij} = \varepsilon_0 \delta_{ij} + 6\bar{n} C\, nT \sum_{\alpha=1}^{N} \left\{ \frac{2}{3} \mu \lambda_\alpha \langle \rho_i^\alpha \rho_k^\alpha \rangle - \frac{1}{3} \delta_{ik} \right. \tag{157}$$

$$\left. + \frac{1}{3T} \int_0^\infty \varphi(s) \left(\langle \dot{\rho}_i^\alpha (t-s) \rho_k^\alpha (t) \rangle - w_{il}(t-s) \langle \rho_l^\alpha (t-s) \rho_k^\alpha (t) \rangle \right) ds \right\}.$$

One admits that the relative permittivity tensor of the system is determined by the mean orientation of the segments, so that we consider expression (157) to be equivalent to relation (151) and, at comparison, can obtain the expression for the mean orientation of segments of macromolecules in an entangled system

$$\langle e_i e_k \rangle - \frac{1}{3} \delta_{ik} = \frac{3}{5z} \sum_{\alpha=1}^{N} \left\{ \frac{2}{3} \mu \lambda_\alpha \langle \rho_i^\alpha \rho_k^\alpha \rangle - \frac{1}{3} \delta_{ik} \right.$$

$$\left. + \frac{1}{3T} \int_0^\infty \varphi(s) \left(\langle \dot{\rho}_i^\alpha (t-s) \rho_k^\alpha (t) \rangle - w_{il}(t-s) \langle \rho_l^\alpha (t-s) \rho_k^\alpha (t) \rangle \right) ds \right\}.$$

where z is number of segments in a macromolecules. Relaxation equations for the mean orientation can be restored. In linear approximation, assuming that $E/B \ll B$, we have

$$\frac{d\langle e_i e_k \rangle}{dt} = -\frac{1}{\tau} \left(\langle e_i e_k \rangle - \frac{1}{3} \delta_{ik} \right) + \frac{\pi^2}{15} \frac{1}{z} B \frac{\tau^*}{\tau} \gamma_{ik} \tag{158}$$

One can see that, in this approximation, disturbed conformation of macromolecules does not affect the mean orientation of segments which in the steady state, can be found from Eq. (158) as

$$\langle e_i e_k \rangle = \frac{1}{3} \delta_{ik} + \frac{\pi^2}{15} \frac{1}{z} \tau^* B \gamma_{ik} \tag{159}$$

It is contrasted with relation (152) for dilute polymer solutions. However, an independent calculation of the tensor of orientation in non-equilibrium situations is much desirable.

7.2
Optical Birefringence

For the given relative permittivity tensor ε_{jl}, the refractive index n of light in the anisotropic medium can be determined from the relation [120, 121]

$$\varepsilon_{jl}E_l = n^2[E_j - s_j(s \cdot E)]. \tag{160}$$

The value of the refractive index depends on the direction of propagation s and on the direction of the polarisation of the light. In the case of an aniso-tropic system, it is convenient to consider particular cases for which the Eq. (157) determines refractive index.

In the simplest cases, the optical anisotropy of polymer systems is studied under the conditions of simple elongation, when the elongation velocity gradient ν_{11} or, in simple shear, when the velocity gradient $\nu_{12} \neq 0$ is given. One frequently deals with the linear effects of anisotropy which are induced by oscillatory velocity gradients or by oscillatory strains

$$u_{ik}(t) = -i\omega\gamma_{ik}(t) \sim e^{-i\omega t}.$$

In this case, it is convenient to characterise the behaviour of the system by the dynamo-optical coefficient $S(\omega) = S'(\omega) + iS''(\omega)$ due to Lodge and Schrag [122] or by the strain-optical coefficient $O(\omega) = O'(\omega) - iO''(\omega)$ due to Inoue et al [123]. These quantities are introduced by relations

$$\varepsilon_{ik} = \varepsilon_0\delta_{ik} + 4\bar{n}S(\omega)\gamma_{ik}$$
$$\varepsilon_{ik} = \varepsilon_0\delta_{ik} + 4\bar{n}O(\omega)u_{ik}. \tag{161}$$

It is easy to find, from previous formulae, that the components of dynamic characteristics are connected by relations

$$O'(\omega) = \omega S''(\omega), \quad O''(\omega) = \omega S'(\omega). \tag{162}$$

Relations (161) are quite similar to the definitions of dynamic viscosity $\eta(\omega)$ and dynamic modulus $G(\omega)$, so these relations are similar to the relations between the components of dynamic modulus and dynamic viscosity (Sect. 6.2).

Dynamo-optical and strain-optical coefficients can be estimated from mea-surements of birefringence Δn under elongational flow or shear flow corre-spondingly

$$\Delta n = 3S(\omega)\nu_{11} = 3O(\omega)u_{11}$$
$$\Delta n = 2S(\omega)\nu_{12} = 2O(\omega)u_{12}.$$

Methods for the experimental estimation of birefringence can be found in the monograph by Tsvetkov et al [117], Janeshitz-Kriegl [118] and in papers by Lodge and Schrag [122], Inoue et al [123] and Kannon and Kornfield [119].

Note that a frequency-dependent stress-optical coefficient $C(\omega)$ can be calculated as ratios

$$C(\omega) = \frac{S(\omega)}{\eta(\omega)} = \frac{O(\omega)}{G(\omega)}.$$

The initial value of $C(\omega)$ is determined by the anisotropy of the macromolecules and does not depend on whether the polymer solution is dilute or concentrated. So, it is an important characteristic of a polymer system.

7.3
Oscillatory Birefringence of Concentrated Solutions

The expression for the strain-optical coefficient $O(\omega)$ for a system of weakly-coupled macromolecules is quite similar to the expression for the dynamic modulus, if the stress-optical coefficient C does not depend neither on frequency nor on the relaxation branch. In this case components of the strain-optical coefficient can be calculated according to formula (131) and have the same form as the components of the dynamic modulus, which are shown in Fig. 3. However, to explain experimental data [123, 124], we must admit that the stress-optical coefficient C depends either on frequency or on the relaxation branch. So as the different relaxation branches are assumably connected with different types of motion, one ought to ascribe different values of the stress-optical coefficient to contributions from different relaxation branches, and the expression for the strain-optical coefficient aquires the following form

$$O(\omega) = nT \sum_{a=1}^{5} \sum_{\alpha=1}^{N} (-i\omega) \frac{C_a \, p_\alpha^{(a)} \tau_\alpha^{(a)}}{1 - i\omega\tau_\alpha^{(a)}} \tag{163}$$

where the times of relaxation and the corresponding weights are given in Sect. 6.2. The stress-optical coefficients are proportional to the polarisability of the structural units of the macromolecule, which can be different for different types of motion of the chain [125]. The strain-optical coefficient of concentrated solutions depends on the dimensionless frequency $\tau^*\omega$ and on the dimensionless parameters

$$O(\omega) = nT f(C_1, C_2, C_3, C_4, C_5, B, \chi, \tau^*\omega).$$

These relations are valid for $\chi \ll 1$, $B \gg 1$. It is determined by five relaxation branches, while the first conformational branch, which is an analogue of the reptation branch, does not contribute practically to the strain-optical coefficient as well to the dynamic modulus. So that it is not essential that this branch is not correctly described in the linear theory.

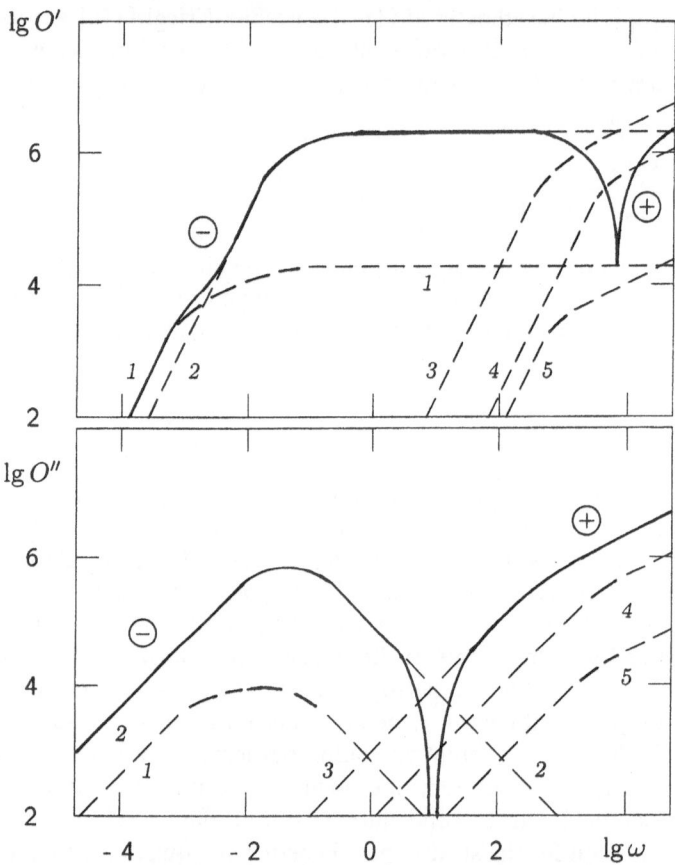

Fig. 4. Strain-optical coefficient of polystyrene. The theoretical dependencies were calculated for polystyrene studied by Onogi et al [97] (see Fig. 3). The separate contributions from relaxation branches are shown by *dashed curves: 1* – conformational branch; *2* – orientation or viscoelastic branch; *3, 4, 5* – glassy branches. The values of parameters are: $B = 3000$, $E = 20,000$, $\chi = 0.08$, $\tau^* = 5 \times 10^{-5}$ s, $nT = 1.7 \times 10^4$ Pa. The stress-optical coefficient is taken different for different relaxation branches: $C_1 = C_2 = -1$, $C_3 = C_4 = C_5 = 0.1$. The characteristic features of the dependence of the strain-optical coefficient on frequency reminds us the empirical ones discovered by Inoue et al [123] for polystyrene with longer macromolecules

Expression (163) allows us to describe different types of the frequency dependence of the strain-optical coefficient which were discovered by Inoue et al [123] and Okomoto et al [124]. As an example a frequency dependence of the real and imaginary parts of the strain-optical coefficient, together with the contributions from different relaxation branches, is shown in Fig. 4. One can notice that the plot on Fig. 4 resembles the empirical picture for polymers discovered by Okomoto et al [124]. So, the mesoscopic theory gives an alternative explanation to the "curious behaviour" of the strain-optical coefficient of polymer solutions and melts. We cannot discuss comparison between the experimental and theoretical curves any more, because it is an illustration of a phenomenon which ought to be investigated carefully.

In application to very concentrated solutions, the optical anisotropy provides the important means for the investigation of slow relaxation processes. It is important to confirm whether or not there is any deviations from the stress-optical law in the low-frequency region for a polymer melt with different lengths of macromolecules. In fact, this is the most important thing to understand the details of the slow relaxation behaviour of macromolecules in concentrated solutions and melts. The evidence can be decisive for understanding the mechanism of the relaxation.

8
Conclusion

The results from recent decades allow us to describe a picture of thermal motion of long macromolecules in a system of entangled macromolecules. The basic picture is, of coarse, a picture of thermal rotational movement of the interacting rigid segments connected in chains – Kuhn – Kramers chains. One can refer to this model as to a microscopic model. In the simplest case (linear macromolecules, see Sect. 3.3), the tensor of the mean orientation $\langle e_i e_j \rangle$ of all (independently of the position in the chain) segments can be introduced, so that the stress tensor and the relative optical permittivity tensor can be expressed through mean orientation as

$$\sigma_{ik} = -p\delta_{ik} + 5nzT\left(\langle e_i e_k \rangle - \frac{1}{3}\delta_{ik} \right),$$

$$\varepsilon_{ij} = \varepsilon_0 \delta_{ij} + \frac{4\pi}{9}nz(\varepsilon_0 + 2)^2 \Delta\alpha \left(\langle e_i e_k \rangle - \frac{1}{3}\delta_{ik} \right)$$

where n is number density of chains in the system, z is a number of segments in a chain; the explanation of other symbols can be found on the previous pages. The relaxation of the tensor of mean orientation of segments $\langle e_i e_j \rangle$ to its equilibrium value is determined by direct rotation of the segments and by slithering of macromolecule along its length contour (reptation) and there are consequently two mechanisms which determine the relaxation of the mean orientation of the segments: rotation of segments and reptation of macromolecule along its length contour. The best way to determine relaxation equation for the tensor of the mean orientation $\langle e_i e_j \rangle$ is to consider the problem of dynamics of the interacting rigid segments connected in chains – the problem which is apparently not solved yet. However, a long flexible macromolecule can be also considered as a chain of coupled Brownian particles, so that the system of entangled macromolecules can be imagined as a suspension of weakly interacting Brownian particles immersed in a viscoelastic liquid. This is a mesoscopic picture of the system which allows us to obtain the results easier. In the first approximation with respect to velocity gradient, the relaxation equation for the mean orientation of the segments has the form

$$\frac{d\langle e_i e_k\rangle}{dt} = -\frac{1}{\tau}\left(\langle e_i e_k\rangle - \frac{1}{3}\delta_{ik}\right) + \frac{\pi^2}{15}\frac{1}{z}B\frac{\tau^*}{\tau}\gamma_{ik}$$

In this approximation, large-scale conformation of the macromolecule, which is controlled by reptation motion of the macromolecule, does not affect the mean orientation of the segments. The reptation relaxation affects only non-linear terms in an expansion of the tensor of mean orientation with respect to velocity gradient.[1] Apart from this, the mesoscopic analysis shows that motion of any Brownian particle of the chain in the system of entangled macromolecules is confined. The particle does not move more than ξ during the times $t < \tau$. For this time of observation, the large-scale conformation of the macromolecule is frozen, but the small-scale motion of the particles confined to the scale ξ can take place, and the macromolecule, indeed, can be considered to be in a "tube" with radius ξ. The macromolecule wobbles around in the tube-like region, remaining near its initial position for some time (a time of localisation) which is the larger the longer the macromolecule is. A very long macromolecule appears, in fact, to behave exactly as if confined in a tube, though no other restrictions than mesoscopic dynamic Eq. (41) exist. The picture of the thermal motion of macromolecules appears to be consistent with description of reptating macromolecules.

Localisation of a macromolecule in the tube was assumed by Edwards [7] and by de Gennes [8] – it was a really very important idea. It was natural then to introduce the reptation mechanism of motion of the macromolecule inside the tube and to try to reduce all relaxation phenomena to the reptation relaxation of macromolecule along the tube. The subsequent analysis in the frame of the formal mesoscopic theory justifies the intuitive introduction of an internal intermediate length – a tube – into the theory, though defines the radius of the tube more precisely in terms of the fundamental parameters (Sect. 5). The analysis justifies the very existence of reptation mobility but discovers that the role of the reptation mechanism in relaxation phenomena was exaggerated. The reptation relaxation exists but practically does not affect linear viscoelasticity and optical birefringence (see Sects. 6 and 7). However, the phenomena of diffusion of long macromolecules cannot be apparently understood without taking into account the reptation mobility. In this case, the reptation mechanism of displacement predominates (Sect. 5). The analysis of the non-linear effects of viscoelastisity [129] also confirms that the reptation relaxation have to be included in consideration to explain correctly the observed dependencies on molecular weight of polymer. The mesoscopic approach gives an amazingly consistent picture of the different relaxation phenomena in very

---

1 That is why the numerous efforts to find the 3.4-index law for viscosity coefficient of linear polymers in frame of the reptation-tube model were doomed to fail and have failed during the last twenty years.

concentrated solutions and melts of linear polymers. It is not surprising: the mesoscopic approach is a kind of phenomenological approach which permits the formulation of overall results regardless of the extent to which the mechanism of a particular effect is understood. In fact, many questions are left unanswered. The need for quantitative comparison with observation, which is the ultimate test of any physical theory, in this case led to the formalism and later to its interpretation in microscopic terms. One can anticipate that the more detailed theories could be very helpful in understanding the thermal behaviour of macromolecules and explaining the introduced mesoscopic parameters. A theory of a deeper level based on the heuristic model of rigid, connected in chains and interacting with each other, segments, we can say 'microscopic' theory as compared to the exploited mesoscopic approach, should give us a description of a few relaxation branches, including orientation and reptation branches. The mesoscopic theory helps us to formulate the correct answer for the problem.

On the other hand, the mesoscopic approach allows us to give justification to some phenomenological constitutive equations [126, 127]. It appears that the correct description of the non-linear effects of viscoelasticity of concentrated solutions and melts of polymers must take into account two (more exactly: two sets of) coupled relaxation processes: orientational and conformational (reptational). The situation appears to be rather complicated, so that the correct constitutive relation has been discovered neither by phenomenological methods of analysis nor by microscopic theories, though there are very good approximations to constitutive relation discussed in monograph [129]. Although the microscopic theory remains to be the real foundation of the theory of relaxation phenomena in polymer systems, the mesoscopic approach has and will not lose its value. It will help to formulate constitutive relation for polymers of different architecture. The information about the microstructure and microdynamics of the material can be incorporated in the form of constitutive relation, thus, allowing to relate different non-linear effects of viscoelasticity to the composition and chemical structure of polymer liquid.

Acknowledgements. I should like gratefully to note that at different times Yu. A. Altukhov, V. B. Erenburg, V. L. Grebnev, Yu. K. Kokorin, N .P. Kruchinin, G. V. Pyshnograi, Yu. V. Tolstobrov, G. G.Tonkikh, A. A. Tskhai, V. S. Volkov and V. E. Zgaevskii participated in the investigations of the problems and in the discussions of the results. I thank them for their helpful collaboration.

It is my great pleasure to acknowledge my indebtedness to Sir Sam Edwards who has kindly read the original manuscript (in fact, a wider version) of the work. His comments and, especially, conversations with him in Cambridge in May 1998 were very useful for me.

Appendices

A
Resistance Force for a Particle in a Viscoelastic Fluid

Let us find the resistance force acting on a spherical particle of radius a which moves slowly with velocity u in an incompressible viscoelastic fluid. It means that the Reynolds number of the problem is small, the convective terms are negligibly small, and the equations of fluid motion are

$$\rho\frac{\partial v_i}{\partial t} = \frac{\partial \sigma_{ij}}{\partial x_j}, \quad \frac{\partial v_i}{\partial x_i} = 0,$$

$$\sigma_{ij} = -p\delta_{ij} + 2\int_0^\infty \eta(s)\gamma_{ij}(t-s)ds, \quad \gamma_{ij} = \frac{1}{2}\left(\frac{\partial v_i}{\partial x_j} + \frac{\partial v_j}{\partial x_i}\right) \tag{A.1}$$

where ρ is a constant density, $v = v(x,t)$ is the velocity of the liquid and σ_i is the density of the outer forces.

A fading memory function $\eta(s)$ can be represented as a sum of exponential functions

$$\eta(s) = \sum_\alpha \frac{\eta_\alpha}{\tau_\alpha}\exp\left(-\frac{t}{\tau_\alpha}\right).$$

The coefficients of partial viscosity η_α and relaxation times τ_α are the characteristics of the liquid.

It is convenient to apply the Fourier transforms of the variables

$$v_i(\omega) = \int_{-\infty}^\infty v_i(t)e^{i\omega t}dt$$

$$p(\omega) = \int_{-\infty}^\infty p(t)e^{i\omega t}dt.$$

So, the equations of motion (A.1) take the form

$$i\omega\rho v_i(\omega) + \frac{\partial p(\omega)}{\partial x_j} = \eta[\omega]\frac{\partial^2 v_i}{\partial x_i\partial x_j}, \quad \frac{\partial v_i}{\partial x_i} = 0, \tag{A.2}$$

where

$$\eta[\omega] = \int_0^\infty \eta(t)e^{i\omega t}dt.$$

It can be easily seen that, if we consider $\eta[\omega]$ as a constant viscosity coefficient, equations (A.2) are identical to the equations of motion of a viscous fluid. As far as a solution of the problem of the motion of a sphere in a viscous fluid is known [128], the resistance force in our problem is written as follows

$$F(\omega) = -6\pi a\eta[\omega]u(\omega)$$

or

$$F(t) = -6\pi a \int_0^\infty \eta(s) u(t - s) ds. \tag{A.3}$$

These results are valid at an arbitrary memory function $\eta(s)$ which can be represented as the sum of exponential functions. In the simplest case

$$\eta(s) = \frac{\eta}{\tau} \exp\left(-\frac{t}{\tau}\right), \qquad \eta[\omega] = \frac{\eta}{1 - i\omega\tau}, \tag{A.4}$$

$$F(t) = -\frac{\zeta}{\tau} \int_0^\infty \exp\left(-\frac{s}{\tau}\right) u(t - s) ds, \tag{A.5}$$

where $\zeta = 6\pi a \eta$ is the friction coefficient of a sphere in a viscous fluid.

B
Resistance Coefficient for a Particle in a Non-Local Fluid

We consider the viscous liquid to be incompressible and the motion of the particle to be slow. It means that the Reynolds number of the problem is small, the convective terms are negligibly small, and the equations of motion of the fluid can be written as follows

$$\rho \frac{\partial v_i}{\partial t} = \frac{\partial \sigma_{ij}}{\partial x_j} + \sigma_i, \qquad \frac{\partial v_i}{\partial x_i} = 0,$$

$$\sigma_{ij}(r) = -p\delta_{ij} + 2 \int \eta(r - r') \gamma_{ij}(r') dr', \qquad \gamma_{ij} = \frac{1}{2}\left(\frac{\partial v_i}{\partial x_j} + \frac{\partial v_j}{\partial x_i}\right) \tag{B.1}$$

where ρ is a constant density, $v = v(x, t)$ is the velocity of the liquid and σ_i is the density of the outer forces. The stress tensor σ_{ij} defines non-local incompressible viscous fluid and contains a decreasing influence function $\eta(r)$, which can be represented as a sum of exponential functions. Integrating over the whole volume is assumed in (B.1).

The motion of a spherical particle in a non-local fluid was considered by Pokrovskii and Pyshnograi [61]. We reproduce the calculation of the resistant coefficient here.

As well as for the local viscous fluid [128], we consider the spherical particle to be immovable and to be situated at the beginning of the co-ordinate frame, so that the flux of the fluid moves around the particle with constant velocity u at infinity.

The equation of motion of the fluid takes the form

$$\frac{\partial}{\partial r_j} \sigma_{ij} = 0 \quad \text{at} \quad |r| > a. \tag{B.2}$$

It is convenient to rewrite the equation of motion as follows

$$\frac{\partial}{\partial r_j}\sigma_{ij} = -f_i(\mathbf{r}).\tag{B.3}$$

Here an induced force $f_i(\mathbf{r})$ is introduced such that equation (B.2) can be determined for all values of the variable $\mathbf{r}$. We shall assume that $f_i(\mathbf{r}) = 0$ for $|r| > a$.

The force acting on the particle can be calculated by integrating over the surface of the sphere or over the volume of the sphere.

$$F_i = -\int_S \sigma_{ij}(\mathbf{r})n_j dS = -\int_V \frac{\partial}{\partial r_j}\sigma_{ij}d\mathbf{r}.$$

Taking equation (B.3) into account, the expression for the force can be re-written as

$$F_i = \int_V f_i(\mathbf{r})d\mathbf{r}.\tag{B.4}$$

Then, we turn to the Fourier transforms of the quantities which can be defined, for example, as

$$f_i(\mathbf{k}) = \int \exp(-i\mathbf{k}\mathbf{r})f_i(\mathbf{r})d\mathbf{r}.$$

So the equations of motion (B.1) and (B.3) for a non-local fluid take the form

$$ip(\mathbf{k})k_i + k_j\eta(\mathbf{k})\left(v_i(\mathbf{k})k_j + v_j(\mathbf{k})k_i\right) = f_i(\mathbf{k}), \qquad k_i v_i(\mathbf{k}) = 0,$$

or

$$k^2\eta(\mathbf{k})v_i(\mathbf{k}) = -ip(\mathbf{k})k_i + f_i(\mathbf{k}), \qquad k_i v_i(\mathbf{k}) = 0.$$

The pressure in the last relations can be excluded so that we have an equation for the Fourier transform of velocity

$$k^2\eta(\mathbf{k})v_i(\mathbf{k}) = \left(\delta_{ij} - \frac{k_i k_j}{k^2}\right)f_j(\mathbf{k}),$$

which has a solution

$$v_i(\mathbf{k}) = \frac{1}{k^2\eta(\mathbf{k})}\left(\delta_{ij} - \frac{k_i k_j}{k^2}\right)f_j(\mathbf{k}).\tag{B.5}$$

The mean velocity of the fluid taken over the surface of the sphere is equal to the velocity of the sphere. In the system of co-ordinates, where the liquid is immovable at infinity we have

$$\frac{1}{4\pi a^2}\int v_i(\mathbf{r})\delta(r-a)d\mathbf{r} = -u_i.\tag{B.6}$$

Relation (B.6) is followed by the relation for the Fourier transform of velocity

$$\frac{1}{(2\pi)^3}\int \frac{\sin ka}{ka} v_i(k)\mathrm{d}k = -u_i. \tag{B.7}$$

Then, we can return to expression (B.6) for the velocity transform and can rewrite relation (B.7) as

$$-u_i = \frac{1}{(2\pi)^3}\int \left(\delta_{ij} - \frac{k_i k_j}{k^2}\right)\frac{\sin ka}{\eta(k)k^3 a} f_i(k)\mathrm{d}k.$$

To calculate the integral, it is convenient to refer to polar co-ordinates and write

$$-u_i = \frac{1}{3\pi a}\frac{3}{8\pi^2}\int (\delta_{ij} - \Omega_i\omega_j)\mathrm{d}\Omega \int_0^\infty \frac{\sin ka}{k}\frac{f_i(k\Omega)}{\eta(k\Omega)}\mathrm{d}k,$$

where $\Omega_i = k_i/k$ is the direct cosine of vector k, and $\mathrm{d}\Omega$ is the differential of the surface of the sphere of unit radius.

So, since the value of the integral does not change when we replace k by $-k$, we can also write

$$-u_i = \frac{1}{6\pi a}\frac{3}{8\pi}\int (\delta_{ij} - \Omega_i\omega_j)\mathrm{d}\Omega\frac{1}{\pi}\int_{-\infty}^{+\infty} \frac{\sin ka}{k}\frac{f_i(k\Omega)}{\eta(k\Omega)}\mathrm{d}k.$$

The last integral can be calculated with the use of the Cauchy theorem about integral values. It results in

$$-u_i = \frac{1}{6\pi a}\frac{1}{8\pi}\int (\delta_{ij} - \Omega_i\omega_j)\frac{f_i(k=0)}{\eta(k=0)}\mathrm{d}\Omega$$

or, eventually,

$$-u_i = \frac{1}{6\pi\eta(k=0)a} f_i(k=0).$$

So, since $f_i(k=0) = F_i$, we obtain the following formula for the force acting on a spherical particle in a non-local viscous fluid

$$F_i = -6\pi\eta(k=0)a\, u_i = -6\pi a\int \eta(r)\mathrm{d}r\, u_i. \tag{B.8}$$

C
Function of Instantaneous Relaxation

To obtain the correct expression for correlation functions when limits are approached, it is convenient to use the function of a non-negative argument

$$R(t) = \lim_{\tau\to 0} e^{-t/\tau} = \begin{cases} 1 & t=0 \\ 0 & t>0 \end{cases}. \tag{C.1}$$

The derivative of the function of instant relaxation is expressed in the delta-function

$$\dot{R}(t) = -2\delta(t).$$ (C.2)

Indeed we can find, by calculation, that

$$\dot{R}(t) = \lim_{\tau \to 0}\left(-\frac{1}{\tau}e^{-t/\tau}\right) = \begin{cases} -\infty & t = 0 \\ 0 & t > 0 \end{cases}.$$ (C.3)

The last expression means that

$$\dot{R}(t) = -k\delta(t).$$ (C.4)

To find the proportionality coefficient, we integrate (C.4) over time. The result confirms relation (C.2).

References

1. Kargin V A, Slonimskii G L (1948) Dokl Akad Nauk SSSR (in Russian) 62:239
2. Rouse P E (1953) J Chem Phys 21:1272
3. Ferry J D (1990) Macromolecules 24:5237
4. Cerf R (1958) J Phys Radium 19:122
5. Zimm B H (1956) J Chem Phys 24:269
6. Peterlin A (1967) J Polymer Sci : A - 2 5:179
7. Edwards S F (1967) Proc Phys Soc 92:9
8. De Gennes P G (1971) J Chem Phys 55:572
9. Doi M, Edwards S F (1986) The theory of polymer dynamics. Oxford University Press, Oxford
10. Edwards S F, Grant J W V (1973) J Phys A: Math Nucl Gen 6:1169
11. Pokrovskii V N (1992) Soviet Phys - Uspekhi 35:384
12. Birshtein T M, Ptitsyn O B (1966) Conformations of macromolecules. Interscience Publishers, New York
13. Flory P J (1969) Statistics of chain molecules. Interscience Publishers, New York
14. Landau L D, Lifshitz E M (1969) Statistical physics, 2nd edn. Pergamon, Oxford
15. Kuhn W (1934) Kolloid Zeitschrift 68:2
16. Mazars M (1999) J Phys A: Math Gen 32:1841
17. Dean P (1967) J Inst Maths Applics No 3:98
18. Flory P J (1953) Principles of polymer chemistry. Cornell University Press, New York
19. De Gennes P G (1979) Scaling concepts in polymer physics. Cornell University Press, New York
20. Alkhimov V I (1991) Soviet Phys - Uspekhi 34:804
21. Edwards S F (1965) Proc Phys Soc 85:613
22. Gabay M, Garel T (1978) J Phys Lett 39:L123
23. Valleau J P (1996) J Chem Phys 104:3071
24. Oono Y, Ohta T, Freed K F (1981) J Chem Phys 74:6458
25. Grassberger P, Hegger R (1996) J Phys A: Math Gen 29:279
26. Yong C W, Clarke J H R, Freire J J, Bishop M (1996) J Chem Phys 105:9666
27. Grossberg A Yu, Khokhlov A R (1994) Statistical physics of macromolecules. AIP Press, Ithaca, New York
28. Polverary M, Van de Ven T G M (1996) J Phys Chem 100:13687
29. Des Cloizeaux J, Jannink G (1990) Polymers in solution: Their modelling and structure. Oxford University Press, Oxford

30. Maconachie A, Richards R W (1978) Polymer 19:739
31. Higgins J S, Benoit H C (1994) Polymers and neutron scattering. Oxford University Press, New York
32. Koniaris K, Muthukumar M (1995) J Chem Phys 103:7136
33. Kholodenko A L, Vilgis T A (1998) Physics Reports 298:251
34. Lodge A S (1956) Trans Faraday Soc 52:120
35. Chandrasekhar S (1943) Rev Modern Phys 15:1
36. Gardiner C W (1983) Handbook of stochastic methods for physics, chemistry and the natural sciences. Springer, Berlin Heidelberg New York
37. Kuhn W, Kuhn H (1945) Helv Chim Acta 28:1533
38. Adelman S A, Freed K F (1977) J Chem Phys 67:1380
39. Dasbach (McMahon) T P, Manke C W, Williams M C (1992) J Phys Chem 96:4118
40. De Gennes P G (1977) J Chem Phys 66:5825
41. MacInnes D A (1977) J Polymer Sci : Polymer Phys Ed 15:465
42. MacInnes D A (1977) J Polymer Sci : Polymer Phys Ed 15:657
42. Peterlin A (1972) Polymer Lett 10:101
44. Rabin Y, Öttinger H Ch (1990) Europhys Lett 13:423
45. Pokrovskii V N (1994) Phys - Uspekhi 37:375
46. Kirkwood J G, Riseman J (1948) J Chem Phys 16:565
47. Curtiss C F, Bird R B (1981) J Chem Phys 74:2016
48. Wilson J D, Loring R F (1993) J Chem Phys 99:7150
49. Herman M F (1990) J Chem Phys 92:2043
50. Herman M F (1995) J Chem Phys 103:4324
51. Edwards S F, Freed K F (1974) J Chem Phys 61:1189
52. Vilgis T A, Genz U (1994) J Phys I 4:1411
53. Born M, Huang K (1954) Dynamical theory of crystal lattice. Oxford University Press, London
54. Schweizer K S (1989) J Chem Phys 91:5802
55. Pokrovskii V N, Volkov V S (1978) Polymer Sci USSR 20:288
56. Pokrovskii V N, Volkov V S (1978) Polymer Sci USSR 20:3029
57. Freed K F, Edwards S F (1974) J Chem Phys 61:3626
58. Freed K F, Edwards S F (1975) J Chem Phys 62:4032
59. Pokrovskii V N, Kokorin Yu K (1984) Vysokomolek Soedin (in Russian) B26:573
60. Pokrovskii V N, Kokorin Yu A (1985) Vysokomolek Soedin (in Russian) B27:794
61. Pokrovskii V N, Pyshnograi G V (1988) Vysokomolek Soedin (in Russian) B30:35
62. Bueche F (1956) J Chem Phys 25:599
63. Doi M, Edwards S F (1978) J Chem Soc : Faraday Trans II 74:1789
64. Des Cloizeaux J (1990) Macromolecules 23:4678
65. Des Cloizeaux J (1992) Macromolecules 25:835
66. Öttinger H Ch (1994) Phys Rev E 50:4891
67. Mead D W, Leal L G (1995) Rheol Acta 34:339
68. Mead D W, Yavich D, Leal L G (1995) Rheol Acta 34:360
69. Chatterjee A P, Geissler Ph L, Loring R F (1996) J Chem Phys 104:5284
70. Milner S T, McLeish T C B (1998) Phys Rev Letters 81:725
71. Riande E, Siaz E (1992) Dipole moments and birefringence of polymer. Prentice Hall, Englewood Cliffs, New Jersey
72. Adachi K, Kotaka T (1993) Prog Polymer Sci 18:585
73. Imanishi Y, Adachi K, Kotaka T (1988) J Chem Phys 89:7685
74. Fodor J S, Hill D A (1994) J Phys Chem 98:7674
75. Van Meerwall E, Grigsby J, Tomich D, Van Antverp R (1982) J Polymer Sci : Polymer Phys Ed 20:1037
76. Fleisher G, Appel D (1995) Macromolecules 28:7281
77. Kokorin Yu K, Pokrovskii V N (1990) Polymer Sci USSR 32:2532

78. Kokorin Yu K, Pokrovskii V N (1993) Int J Polym Mater 20:223
79. Klein J (1986) Macromolecules 19:105
80. Seggern J von, Klotz S, Cantow H J (1991) Macromolecules 24:3300
81. Silescu H (1991) J Non-Cryst Solids 131 - 133, part II:593
82. Daoud M, Cotton J P, Farnoux B et al (1975) Macromolecules 8:804
83. Higgins J S, Roots J E (1985) J Chem Soc : Faraday Trans II 81:757
84. Richter D, Farago B, Fetters L J, Huang J S, Ewen B, Lartique C (1990) Phys Rev Lett 64:1389
85. Hansen J P, McDonald J R (1976) Theory of simple liquids. Academic Press, London
86. Ronca G (1983) J Chem Phys 79:1031
87. Des Cloizeaux J (1993) J Phys I France 3:1523
88. Richter D, Farago B, Butera R, Fetters L J, Huang J S, Ewen B (1993) Macromolecules 26:795
89. Ewen B, Richter D (1995) Macrololecul Symp 90:131
90. Ewen B, Maschke U, Richter D, Farago B (1994) Acta Polymer 45:143
91. Genz U, Vilgis T A (1994) J Chem Phys 101:7101
92. Kholodenko A L (1996) Macromol Theory Simul 5:1031
93. Curtiss C F, Bird R B (1981b) J Chem Phys 74:2026
94. Wittmann H -P, Fredrickson G H (1994) J Phys I France 4:1791
95. Rice S A, Gray P (1965) Statistical mechanics of simple liquids. John Wiley, New York
96. Gray P (1968) The kinetic theory of transport phenomena in simple liquids. In: Physics of simple liquids. North-Holland Publishing Company, Amsterdam, p 507
97. Onogi S, Masuda T, Kitagawa K (1970) Macromolecules 3:109
98. Baumgaertel M, Schausberger A, Winter H H (1990) Rheol Acta 29:400
99. Baumgaertel M, De Rosa M E, Machado J, Masse M, Winter H H (1992) Rheol Acta 31:75
100. Ferry J D (1980) Viscoelastic properties of polymers, 3rd ed. Wiley, London
101. Aharoni S M (1983) Macromolecules 16:1722
102. Aharoni S M (1986) Macromolecules 19:426
103. Berry G C, Fox T G (1968) Advance Polymer Sci 5:261
104. Poh B T, Ong B T (1984) European Polymer J 20:975
105. Takahashi Y, Isono Y, Noda I, Nagasawa M (1985) Macromolecules 18:1002
106. Phillies G D J (1995) Macromolecules 28:8198
107. Yanovski Yu G, Vinogradov G V, Ivanova L N (1982) Viscoelasticity of blends of narrow molecular mass distribution polymers. In: Novyje aspecty v reologii polimerov (New Items in Polymer Rheology, in Russian) part 1. INKHS AN USSR, Moscow, p 80
108. Jackson J K, Winter H (1995) Macromolecules 28:3146
109. Watanabe H, Kotaka T (1984) Macromolecules 17:2316
110. Watanabe H, Sakamoto T, Kotaka T (1985) Macromolecules 18:1008
111. Van Vleck J H (1932) The theory of electric and magnetic susceptibilities. Oxford University Press, London
112. Fröhlich H (1958) Theory of dielectrics, dielectric constant and dielectric loss, 2nd edn. Clarendon Press, Oxford
113. Gotlib Yu Ya, Svetlov Yu E (1964) Vysokomolek soedin (in Russian) 6:1591
114. Gotlib Yu Ya, Svetlov Yu E (1964) Vysokomolek soedin (in Russian) 6:771
115. Gotlib Yu Ya (1964) Vysokomolek soedin (in Russian) 6:389
116. Zgaevskii V E, Pokrovskii V N (1970) Zh Prikl Spektrosk (in Russian) 12:312
117. Tsvetkov V N, Eskin V E, Frenkel S Ya (1964) Struktura makromolekul v rastvorakh (Structure of macromolecules in solutions, in Russian). Nauka, Moscow

118. Janeschitz-Kriegl H (1983) Polymer melt rheology and flow birefringence. Springer, Berlin Heidelberg New York
119. Kannon R M, Kornfield J A (1994) J Rheol 38:1127
120. Born M, Wolf E (1970) Principles of optics. Electromagnetic theory of propagation, interference and deffraction of light, 4th edn. Pergamon, Oxford
121. Landau L D, Lifshitz E M, Pitaevskii L P (1987) Electrodynamics of continuous media, 2nd edn. Pergamon, Oxford
122. Lodge T P, Schrag J L (1984) Macromolecules 17:352
123. Inoue T, Okamoto H, Osaki K (1991) Macromolecules 24:5670
124. Okamoto H, Inoue T, Osaki K (1995) J Polym Sci B Polym Phys 33:1409
125. Gao J, Weiner J H (1994) Macromolecules 27:1201
126. Altukhov Yu A, Pyshnograi G V (1996) Polymer Sci A 38:766
127. Pokrovskii V N, Altukhov Yu A, Pyshnograi G V (1998) J Non-Newtonian Fluid Mech 76:153
128. Landau L D, Lifshitz E M (1987) Fluid mechanics, 2nd edn. Pergamon Press, Oxford
129. Pokrovskii V N (2000) The mesoscopic theory of polymer dynamics. Kluwer Academic Publishers, Dordrecht Boston London

Editor: Prof S. Edwards
Received: June 2000

Author Index Volumes 101–154

Author Index Volumes 1–100 see Volume 100

Subject Index